the warrior women of television

Intersections in Communications and Culture

Cameron McCarthy and Angharad N. Valdivia
General Editors

Vol. 7

PETER LANG
New York • Washington, D.C./Baltimore • Bern
Frankfurt am Main • Berlin • Brussels • Vienna • Oxford

DAWN HEINECKEN

the warrior women of television

A Feminist Cultural Analysis of the New Female Body in Popular Media

PETER LANG
New York • Washington, D.C./Baltimore • Bern
Frankfurt am Main • Berlin • Brussels • Vienna • Oxford

Library of Congress Cataloging-in-Publication Data

Heinecken, Dawn.
The warrior women of television: a feminist cultural analysis of the
new female body in popular media / Dawn Heinecken.
p. cm. — (Intersections in communications and culture; vol. 7)
Includes bibliographical references and index.
1. Women on television. 2. Body, Human, on television.
3. Women heroes. 4. Television broadcasting—
Social aspects. I. Title. II. Series.
PN1992.8.W65 H45 791.43'082—dc21 2002152521
ISBN 0-8204-6219-5
ISSN 1528-610X

Bibliographic information published by **Die Deutsche Bibliothek**.
Die Deutsche Bibliothek lists this publication in the "Deutsche
Nationalbibliografie"; detailed bibliographic data is available
on the Internet at http://dnb.ddb.de/.

Cover design by Lisa Barfield

275 Seventh Avenue, 28th Floor, New York, NY 10001
www.peterlangusa.com

TABLE OF CONTENTS

ACKNOWLEDGMENTS

I would like to express gratitude to several people who helped me complete this project. Among the friends and colleagues to whom I am grateful are Vickie Rutledge Shields, who oversaw the original manuscript on which this book is based, and Nancy Theriot and Lin Billingsley, whose comments and encouragement have been invaluable.

The author and publisher would also like to thank the following for permission to use the following copyrighted material:

Stills from *Buffy the Vampire Slayer*. Used with permission by Twentieth Century Fox Television. All Rights Reserved.

Stills from *La Femme Nikita*. Used with permission by Warner Brothers. All Rights Reserved.

Stills from MTV's *Aeon Flux*. Used with permission by MTV: Music Television. © 1998 MTV Networks. All Rights Reserved. MTV: Music Television, all related titles, characters and logos are trademarks owned by MTV Networks, a division of Viacom International Inc.

Dawn Heinecken, "Boundary Battles: Heroic Narrative, the Feminine, and MTV's *Aeon Flux*," *Studies in Popular Culture* 24.1 (2001). Reprinted by permission of the publisher. All rights reserved.

Thank you also to the University of Louisville's Committee on Academic Publication for its help in obtaining permissions.

INTRODUCTION

Feminism has taught us that the story both begins and ends with bodies.[1]

From early Westerns to recent films and series like *The A-Team, Rambo, Die Hard* and *Lethal Weapon*, the action hero's body is stained with blood and sweat. The hero emerges victorious after being shot, battered, or tortured in diabolical ways. The marks of violence engraved on his body signify his ability to overcome all physical suffering, yet the cuts and scars also reveal a deeper understanding of his character. It is apparent throughout that it is the hero's indomitable will that drives his body to such extraordinary ends. The hero's taut, heavily muscled body summarizes the genre. At once bleeding and bruised, his chiseled sinews belie the hero's suffering by evoking a "hardness" that exists beyond the physical. The hardness of the hero's body works to define him—as man, as master over his environment.

The action hero has a control over his own body that has historically been denied to women, the weaker, "softer" sex. While heroes forge change in their environments through their hard bodies, the soft bodies of women have seemed to recoil from expansion. It is no surprise then that women have rarely served as action heroes. Even when women have, the emphasis has usually been on their voluptuous bodies rather than their heroic actions. I still remember the pain I felt at hearing my older brother contemptuously refer to Lynda Carter's Wonder Woman as "Wonder Tits." In that one degrading word, his disgust for her softness, her femininity, her lack of heroic grit, was summarized. All her superpowers were for nothing compared to the weakness of her soft body. Even my hero was not exempt from the devaluation of women's bodies that surrounded me. What would it be like to have grown up with female heroes whose bodies resisted this mold?

On the surface, today's female heroes seem to be the answer to my prayers. They are very different from the female heroes of the 1960s or 70s, like those of *Get Christie Love*, *Wonder Woman*, or *Charlie's Angels,* who jiggled across the screen. Today's heroines, like *Buffy the Vampire Slayer*, *La Femme Nikita*, *The PowerPuff Girls*, *Dark Angel*, or *Xena, Warrior Princess* seem harder, stronger, and tougher. In fact, they are often held up as role models for young girls. Yet is the situation really that simple? *Warrior Women* asks the question: What does the current proliferation of female action heroes in television really tell us about contemporary attitudes towards the female hero, the female body, and female power?

Before examining the "meaning" of today's array of female heroes, it is necessary to examine the context from which they emerged. Just as the female hero has been a figure of fun in comparison to the bloodstained, sweaty, but masterful hero, women's bodies have consistently been devalued by a male-dominated culture in real life. This attitude has had a profound effect on the way women live in and experience their bodies.

For example, in the recent film *War Zone* by Maggie Hadleigh-West, Hadleigh-West takes a video camera with her onto the city streets.[2] She turns her camera on men as they whistle and comment on her body, subjecting them to the same kind of scrutiny. The discomfort, anger, and aggression that her victims demonstrate at this reversal are telling. Their anger reveals how thoroughly it is accepted that men dominate public space. Men have the potential to say and do anything to women, but women do not have the same potential power to invade the personal and physical space of men. This means that women's physical freedom is frequently curtailed. How many women do you know who walk alone at night? The fact that women do not have the same access to public space—the same right to walk down the street unmolested—is an example of how women are oppressed by male dominance. When a woman asserts herself in public space, as does Hadleigh-West, she threatens the everyday order of things. She treats men as women are commonly treated. This is why the men in the film react with so much rage at her "invasion" of their privacy.

War Zone demonstrates the link between sight and power. The men whistle, hoot, holler, and subject women to sexual scrutiny. The men's often obvious sexual looks at Hadleigh-West exemplify the fact that many men feel a given right to look. Their gaze is a sign of their social power. In *Ways of Seeing*, John Berger notes that "men act and women appear."[3] Women in our culture have been raised with the awareness that we are most often judged based on our appearance—the appeal of our bodies to men. Our bodies should be sexy, lovely, fragrant, and, above all, passive. We are taught to be objects of the male gaze rather than to look back at and evaluate men in turn. When a woman *looks back at men*, as Hadleigh-West does with her camera, she reverses the gaze and claims male power.

These traditional ideas of the feminine as passive object actually limit the physical activity of the female body. In *Throwing like a Girl*, Iris Marion Young observes the ways in which women use their bodies in day-to-day life. She found that a chief characteristic of women's physical movement is that it takes up the *smallest* possible space when performing an activity—whether running, walking, or sitting.[4] This body language expresses the awareness that women's bodies are to be looked at rather than actively used. A woman should provide a pleasing exterior, but she should never use her body in a way that might disrupt male power. Similarly, studies such as Carole Spitzack's and Susan Bordo's work on women's obsession with weight loss show how mainstream culture and the media urge women to make their bodies small and to take up less space.[5] The less space you take up, the less threatening you are.

"Bitch" is a common name given to women who resist this rule. As Kathleen Rowe has written, women who break the "feminine" mold of smallness and take up space through their physicality, large appearance, or loud voice are considered "unruly" women. Their presence, like Hadleigh-West's, threatens the male dominance of public space and suggests the inversion of patriarchal norms. Because of this the unruly woman is often a figure of ridicule and hatred or represented as monstrous or obscene.[6]

These studies all reveal common themes at the heart of women's experience of the physical self: Women's bodies are valued for the spectacle they provide, but at the same time women's bodies must be small and unthreatening and must never intrude upon public "male" space. One central idea of many of these scholars is that the media play an important role in teaching us this lesson. The media are seen as major instruments of cultural expression. They both reflect and promote a particular belief system, an *ideology*, in which men maintain their social power through the subordination of women. Looking at how popular media like film and television represent gender relationships, women, and the female body can thus tell us much about the governing ideologies of the culture itself.

The connection between patriarchal ideologies and representations of the female body has been at the heart of feminist film studies for over two decades. It has been most emphasized by psychoanalytic film criticism. Developed from Laura Mulvey's groundbreaking article, "Visual Pleasure and Narrative Cinema," these critics show how mainstream movies are filmed and edited and are influenced by deep-seated ideas about masculinity and femininity. As a result, a guiding visual principle of Hollywood films is that women and their bodies are consistently disempowered and contained.

Mulvey's article examines the ways that film structure naturalizes male supremacy through its representation of the feminine. Her argument centers on the idea of "the triple gaze": the gaze of the film's hero, called the subject; the gaze of the camera; and the gaze of the film "spectator" or viewer. In many cases, these gazes are aligned. The audience member, or spectator, watches the film unfold as if through the eyes of the male protagonist, what he and we see is also what the camera sees. The "subject" is thus *more than* the hero. It is a theoretical position or space created by the intersection of all these gazes. This subject position is active and "masculine." Even if the sex of the audience member is female, she must take up the "male" subject position.

In Freudian psychoanalytic theory, the female body is seen as representing a threat to the male's sense of self-completeness. Mulvey, therefore, reasons that cinema soothes male anxiety by either "punishing" the female in the story line or by "fetishizing" the female body. In film, the female body tends to be the recipient of all three gazes, often at the same time, and thus appears extremely passive. Mulvey shows how film techniques like editing, framing, and lighting work to emphasize body parts and fragment the female body. While glamorous and idealized, these images work to disempower the women because they are static and motionless. It is as if female body parts are held captive by gaze of the camera and the spectator. Psychoanalytic film theory thus argues that film reiterates the cultural view of the female body as passive, objectified, and lacking in agency.[7]

Not surprisingly, since Mulvey's article appeared, feminist film criticism has been preoccupied with the search for female agency in film. Recent work in film and television studies suggests that the dominant ideology has begun to incorporate at least some of the values of the feminist movement. These studies find proof of shifting cultural norms in the changing representations of women in mainstream film and television. They argue that the representation of women has changed with time, pointing to the media's growing acceptance of some basic feminist ideas, such as the right to female independence and the right to work.[8]

The recent trend of female action heroes in film and television is one notable change. In *Cracks in the Pedestal*, Philip Green sees their presence as a sign that feminist beliefs and attitudes have at least begun to be incorporated by mainstream films, pointing to a wide range of films and television programming that feature a woman as the hero-subject.[9]

Action adventure films starring women break with traditional cinematic representations of gender because they situate the women in the place formerly occupied by the male hero. Not only do a growing number of films like *The Fifth Element*, *Barb Wire*, and *Tomorrow Never Dies* or television shows like *Prime Suspect* feature women in starring or lead action roles, but, as Philip Green argues, a range of films, such as *China O'Brien* or *The Stranger*, actually allows females to be seen as "the individual."[10] In other words, in these films, the female hero has become "the bearer" of the look—the source of the point of view, narrative agency, and audience identification.

However, media forms, such as film, TV, and advertising, are notorious for the ways in which they adopt and absorb oppositional points of view or modes of expression, reframing them in ways that are beneficial to big business and the ruling corporate class. For example, punk rock was initially a rebellion against mainstream music and capitalist values. When it was picked up and marketed by the mainstream industry, punk broke its own rules, making money for various corporations. In the process, the social resistance at the heart of punk was destroyed. Likewise, Susan Willis and Susan Bordo have shown how the women's fitness movement, which was at first oppositional to a culture that denied female activity, was co-opted by big business and advertisers whose main goal is to make money. Female fitness eventually became yet another demand on women for physical perfection.[11]

It is therefore possible that the media industry may be adopting a "feminist face" by the inclusion of female heroes—giving a nod to the many feminists in the land—while constructing these shows in such a way as to diffuse their feminist potential. Before deciding whether or not these female heroes express a pro-feminist ideological shift in mainstream culture, it is therefore necessary to examine exactly how certain media texts represent female bodies that take up space. *The Warrior Women of Television* investigates the current representation of female action heroes on television to see what they have to say about female bodies and, by extension, female power.

Using a case-study approach, I examine three television shows that are representative of the range of modern female heroes: *La Femme Nikita*, *Aeon Flux*, and *Buffy the Vampire Slayer*. These three shows were chosen for examination for several reasons. First, at the time this book was written, these three series, along with *Xena: Warrior Princess*, were the only television series featuring female heroes at the center of the storyline. As such, they are linchpins, foundational series in the new trend of female heroes on television. The series are also of particular interest because of their mainstream presence on popular cable channels as well as their popular and cult appeal. Cult status—evidenced by the large number of on-line fan sites dedicated to the series—means that these series appeal not only to the "average" audience member but also to a particularly focused and motivated group of people to whom the series speaks deeply. The participatory culture of fans of shows with similar cult appeal, like *Star Trek* or *Highlander*, is testimony to the fact that fans may incorporate ideals or fantasies about the show into their daily lives and personal identities.[12] Examining these three series also gives balance to this study since each series embodies a different approach to the female hero, ranging from comedic (*Buffy*), dramatic (*Nikita*), to animated (*Aeon*) in form.

This book draws from a long history of feminist studies of female bodies, female representation, and the media and connects to recent feminist television and film studies that are concerned with changing media representations of women. Semiotics, "the study of signs," is a way of looking closely at the imagery used in "texts" (films, TV shows, ads). Imagery is not accidental. The "signs" that are used in a text—such as a bouquet of red roses—carry a range of meanings and associations which contribute to our evaluation of the text's "message." A bouquet of red roses conveys associations with love, relationships, romance, etc. However, the relationship of signs to other signs in a text also influences the overall meaning of a text. A bouquet of flowers in the hands of a young child conveys a very different meaning from a bouquet of flowers placed on a grave. This book will analyze the "signs" at play in each of the three series. It will also examine the formal properties—the film techniques like narrative, editing, lighting—that help build interpretations about "the meaning" of the female hero.

Warrior Women pays particular attention to the signs of the body. The term "body" refers not only to the presentation of the body of the actress, the social roles she adopts, or her actions but also includes other considerations. Anne Balsamo describes the body as both a "product and a process." It is a product in that it is a material embodiment of ethnic, racial, and gender identities as well as a performance of personal identity. It is a process in that it is "a way of knowing and marking the world, as well as a way of knowing and marking a 'self'."[13] Body studies are thus connected to epistemology and ontology, our ways of being and understanding ourselves and our place in the world. As such, representations of the body have philosophical as well as political implications.

As Norman Fairclough has observed, the relationship between a text and culture is dialectic, relational. Analysis of texts, therefore, cannot be based on an understanding of the text alone. It needs to be done along with an analysis of wider cultural discourses.[14] Discourse is perhaps best visualized as a web. The web is made up of various strands that are ways of thinking and talking, beliefs, attitudes, and behaviors. Individual strands overlap and overflow, twining in and about, interacting with, and informing each other. The web surrounds us, shaping our ways of thinking and talking in general and about certain issues. Discourse frames our understanding of the world. It sets the pattern for, and limits, what is and is not discussed, what is and is not done or thought about in a society or in a text.

Interpretation of a text needs to be informed by an understanding of how the text fits in amongst larger cultural discourses. This means that not only do we have to look at the context of the culture and historical period in which a text emerges and is watched, but we also have to look at the interplay of texts with other texts that are also part of the larger cultural discourse. When we watch an action film, how we understand it is influenced by several factors. The meanings we take from watching the movie *Star Wars*, for instance, are formed by the time and place in which we live. People who watched *Star Wars* for the first time when it was released in 1977 probably had a very different reaction to it than those who watched it first in the year 2000. The cultural frame has changed so that what we expect from special effects, what we expect from Harrison Ford, who is a now a big star, differs. Our understanding of the film will also be linked to our previous knowledge of the genres of action and of sci-fi, stories of heroic quests, and even the press coverage of a film: Is it "trash" or "highbrow?" What's the gossip about the actors in this particular scene? Likewise, Julie D'Acci has shown how extratextual discourse like magazines and advertisements imply and may even construct a gendered audience.[15]

Accordingly, extratextual materials such as series ads, web sites, magazine articles, and fan fiction are included in this analysis since they may inform interpretations of the texts. Similarly, this book examines the three series through a lens informed by other popular media texts, which similarly help instruct audiences on how to interpret the text. For example, in discussing the representations of the female hero's body, I will be comparing the texts to past representations of male action heroes as well as discussing their connections with popular romance and horror novels.

Warrior Women pays attention to the ways *Buffy, Nikita*, and *Aeon* reflect, extend, or comment upon existing public discourses. Hence, the last chapters of this book connect the major themes and ideas expressed by these series to insights from critical writings on contemporary culture and the body. The final chapters locate these texts amongst the imagery and practices of our postmodern, postindustrial era and discuss some of their implications in regards to issues in mainstream feminism.

Chapter One traces the evolution of feminist media studies and argues for the need to study texts as mediations of cultural concerns. Chapter Two outlines the findings of previous studies about female heroes and discusses the typical imagery of male-centered texts and the ideas this imagery expresses about the body. Chapters Three through Five provide detailed analyses of the three series. Chapter Six discusses the ways in which the female hero mediates contemporary anxiety over the body and the breaking down of both bodily and other boundaries. The Epilogue concludes with a discussion of how the series' emphasis on notions of bodily control and bodily pleasure mediates issues tied to current understandings of female power that is expressed by "postfeminism."

While the findings in this book undoubtedly reflect my personal position as a young, white female, the purpose of my analysis is not to determine one conclusive reading of each text but rather to unravel various threads of meaning to explore the ways that each text *might* be read. This project understands the relationship between media texts and audiences as a process of negotiation. The critic's task is to examine the "multiple and contending meanings embedded in a text as these might appear to particular social subjects" and to examine how meanings intersect with historical and social context.[16] *Warrior Women* addresses the question of what current representations of active female bodies have to say about our cultural attitudes and beliefs regarding bodies and power.

CHAPTER ONE

Feminist Media Theory

Laura Mulvey's 1975 article, "Visual Pleasure and Narrative Cinema," shaped the direction of media studies for years to come. Mulvey argued that "classic" Hollywood films like those of Alfred Hitchcock use lighting, camera angles, editing, and story lines in calculated ways to create specific effects. Story lines, or narratives, are driven by the action of the hero. Films are put together in such a way that when we watch a movie, we see through the hero's eyes as the events of the film world unfold. In other words, the male hero is the subject of the film, the person with whom the audience is encouraged to identify and whose point of view we are supposed to share.[1]

However, women in Hollywood films are in a different position altogether. Their action is rarely central to the narrative. Often their entrance into a story works to slow the story down as, for example, in love scenes that have nothing to do with the larger plot. Mulvey also showed that the way women are filmed reinforces the message that women are passive. Women in films are lit and framed in a way that encourages audiences to look at their bodies. Moreover, since we are already seeing through the eyes of the male hero, we identify with the hero as he looks *at* the woman. She is literally the direct object of both the hero's and the audience's gaze.

Although in most films we are encouraged to see through the hero's eyes as he looks at a woman, we rarely see through the female's eyes, and we never see through her eyes as she looks *back at* the hero. This means that the male is looking at the woman without, in turn, being seen himself. He is in the powerful position of the Peeping Tom, a voyeur.

An example of how the ability to look without being seen is a kind of power is illustrated by the beginning of the recent movie, *Scream*. As the movie opens, a character played by Drew Barrymore has a flirtatious conversation on the phone with a stranger. She does not begin to feel menaced until he is able to describe her appearance and actions. We realize at the same moment she does that this strange man is watching her. The terror of this moment stems from the fact that he can see her, but she cannot see him. He could be anywhere, even inside the house, but she cannot see him to even *begin* to fight.

Most women in Hollywood films are treated a bit like Barrymore's hapless victim; unable to look back at the male character who is watching her, she appears disempowered and weak in contrast. The disempowerment of women occurs in other areas as well; often the story lines "punish" female characters—with women either being beaten or killed, suffering or dying. A key way in which audiences are encouraged to see the woman as passive and weak is through the traditional method of filming a woman in which only one part of her body is seen at a time, as, for example, with shots focusing solely on breasts, legs, or bottom. This technique is called fragmentation, and it is a staple technique in filming everything from music videos to mainstream films. Fragmentation works to draw our attention to these parts so they seem ultra-sexy, but it is also is a way of reducing women to just her parts—she becomes, not a complete individual with her own unique personality, but a thing, an object—just a pair of legs or breasts rather than a human being.

Mulvey's article created some problems for female audience members because it basically set up a no-win situation for women in film. In order for women to enjoy watching Hollywood films, Mulvey essentially argued that female audience members had to identify with the male subject position and either enjoy watching the female characters be sexualized or enjoy watching women be objectified, disempowered, and punished.[2] Her article also suggested that the traditions of filming were reflections of the psyches and deep, unconscious attitudes of both filmmakers and audiences towards women, and because of this, it would be impossible for a female character to serve as the "hero"—the subject of the film. Even if a movie featured a gender reversal and had a female hero or used female-identified point-of-view shots, it would be unlikely that film spectators, living within a sexist culture, would ever interpret a female body as not-object.

Many feminist scholars saw this as a rather joyless, hopeless conclusion. Tania Modleski, Linda Williams, and later Mulvey herself argued that audience members had more choice than this. Modleski, in *Loving with a Vengeance*, examined female media genres such as television soap operas and romance novels. She wrote that not only did soaps and romances deal with specifically "female" concerns, like relationships, but they were composed in such a way audiences could identify with multiple characters.[3]

Modleski also pointed out that while romances and soaps tend to be about a woman solving her problems by falling in love with a man, romances, for example, often simultaneously criticize relationships in which the heroine is seen as subservient or "less than" the male hero. Indeed,

Modleski found that romances were often about the ways in which the "disempowered" heroine enters into a dangerous situation with a threatening male and turns it to her advantage by winning his love. The romance heroine thus is an active subject with power over her environment.

In two well-known articles, Linda Williams likewise argued a female subject within Hollywood films was possible. Williams contended that the film *Stella Dallas* was structured in such a way that it was possible for female audience members to use their "feminine" empathy to identify with far more than one character. In this study, as well as her investigation of *Mildred Pierce*, Williams points to the ways that the film's story line presents a contradictory "message." On the one hand, the audience is encouraged to accept the heroine's sacrifice; on the other hand, the film presents this sacrifice as unjust. Thus, even though mainstream films did tend to objectify women, there were also places within certain films that broke with the tradition.[4]

Elizabeth Traube points out that while these scholars found that women's portrayals were not as horrible as initially suggested by Mulvey, they all maintained that the "meaning" of a film could be determined by examining filming techniques. Techniques such as editing, for example, control what the audience sees and when it sees something. It can thus direct the audience into identifying with specific characters or points of view. Further, these studies argued that the reason why films were put together in particular ways and the reason why they appealed to audiences could be explained by psychoanalysis.[5] In other words, films revealed the unconscious thoughts, attitudes, and desires of both filmmakers and audiences.

Yet focusing solely on film structure and psychoanalysis may be misleading. Traube has written that this approach overlooks the relationship between the film to the real, outside world surrounding it. She argued that just as the audience of 1990 differs from the audience of 1950 so does the film's "meaning." In addition, psychoanalysis—which explains individual psychology on the basis of gender—ignores how race, class, and sexuality may influence how an individual sees a film. Psychoanalysis ignores context: Human attitudes, beliefs, and values change over time and depend upon the culture in which they live.

Questions surrounding how different groups of people relate to films and television were addressed by another branch of media studies that appeared in the eighties. Cultural studies, emerging from Marxist scholarship, took a very different approach towards media. Cultural studies scholars argued that previous film theory made the film or "text" too important and implied that audience members were passive, not active, interpreters of what they were viewing. While film theory had tended to see individual audience members as "blank slates," cultural theorists argued that an individual audience member's life and social experiences would inform her interpretation of a text. For example, a poor, white, rural woman would interpret a text quite differently than a wealthy, urban, black woman. Cultural theorists thus argued that it was no longer possible to discuss the "meaning" of a film or TV show without talking to its actual viewers. Audience response research was necessary in order to determine the "meaning" of a text.

Since textual meanings would change depending on the viewer, earlier media research was criticized as examples of how an educated and politically motivated individual interpreted a text rather than how it was "truly" seen by "real" viewers. In contrast, audience studies scholars wanted to understand the ways that diverse social groups interpreted and used media texts in their daily lives. Most noticeable among these were Ien Ang's study of *Dallas* fans, Janice Radway's study of romance readers, David Morley's study of the *Nationwide* audience, and Jackie Stacey's research on women's perceptions of forties' and fifties' film stars, *Star Gazing*. Their research showed that audiences, far from being passive recipients of a message created by filmmakers, are actually very active in their interpretations of what a film or TV show "means."

These studies found that there were three basic ways in which audiences interpreted or "read" media texts. The first was to unquestioningly accept the most apparent meaning of the film put forward by the filmmaker, a "dominant" or "preferred reading." In the second, called a "negotiated reading," audiences often pick and chose how they see a film or TV show—accepting some parts and rejecting others. Finally, a "resistant" or "oppositional reading" was one in which audiences rejected the text entirely as stupid, unbelievable, racist, sexist, etc.

An example of a negotiated reading is found in Janice Radway's study of romance readers.[6] Many romances include scenes in which the heroine is raped or is made submissive to the male hero. Many people assume that romance readers do dominant readings: in other words, romance readers enjoy seeing the heroine humiliated or punished. This is taken as a sign that romance readers are stupid, passive, and conform to traditional ideals of femininity. However, Radway showed that in reality romance fans negotiate with texts. For example, many fans chose to ignore or skip over parts in which the heroine is brutalized. Also, romance readers do not see romances as being all the same. Many fans made distinctions between "bad" romances, which tended to be those in which the heroine is submissive and brutalized, and ideal romances, in which the heroine achieves a mutual relationship with the hero. In addition, Radway took into account what the act of reading itself meant in the lives of romance fans. She found that romance readers used the act of reading as a form of escape—time to themselves when they could ignore the demands of family. Thus, reading romances serves as a way for fans to rebel against their roles as wives and mothers. Far from being passive dupes, romance readers are selective in what they read and express discontent with traditional female roles.

This kind of audience research accomplished several important things. First, it paid attention to the audience and gave credit to human beings for being reasonable, discerning media consumers—not brainwashed dullards. Secondly, audience researchers highlighted the need to take into account *context*—the cultural climate, period in history, etc.—when examining media.

This emphasis on active audiences also led to the rejection of another approach towards film studies, generally referred to as "images of women." This approach made generalized conclusions about the governing attitudes and beliefs of a time based on the researcher's interpretation of Hollywood films. Films, in other words, can be seen as "mirrors" of reality. For example, in her book, *From Reverence to Rape*, Molly Haskell wrote that comedies and musicals of the forties recognized female power, reflective of the fact many women were in the work force during WWII. However, the films of the fifties reflected a cultural backlash towards working women and urged a return to traditional female roles and the cult of domesticity.[7] Haskell and others suggested that in a sexist society, the images of women would remain sexist until social structures changed.

Audience reception studies refuted the "images of women" belief that analysis of movie texts could reveal a *common cultural truth*. This is due to the fact that audience members are not passive but active. Therefore, just as the content of romance novels does not necessarily reflect the views of romance readers, the content of films does not necessarily reflect the views of a unified culture. Indeed, in the postmodern world and the proliferation of numerous subcultures within the United States—subcultures based on region, race, class, age, sexuality, and so forth—it has been increasingly argued that there is no *one* culture. Further, with the range of media now from network TV, cable, video, and films, it is not possible to even talk about a dominant form of media. Thus the media can no longer be seen as a "mirror" of reality, since the question becomes, which media, whose reality?

Despite the importance of the insights gained from audience reception studies, such studies are not without problems. To a certain extent, audience reception argues that the meaning of texts is in the eye of the beholder. Taken too far, audience reception theory suggests that each audience member sees an entirely different film or show than the person sitting next to them. For example, it suggests that certain fans might see Captain Kirk as the hero of *Star Trek*, while others might see Nurse Chapel as the series' central heroic figure. While certainly some individuals might identify with or like Nurse Chapel better than Kirk, to say that the series presents the characters as equals, would involve a complete rejection of much of what we actually *see*—many, many minutes of Kirk ordering his crew about, fighting, loving, saving the day, etc., and very few moments of Nurse Chapel handing Dr. McCoy equipment or swooning over Spock. The form and structure of the series clearly locate Chapel as a minor figure.

In addition, audience reception studies tend to place a lot of importance on audience likes and dislikes. Taking pleasure in reading a romance, for example, can be seen as a sign of "resistance" because the readers are ignoring their husbands, for example. However, a criticism of audience reception studies has been that they focus too much on audience pleasure.[8] By focusing on audience likes and dislikes to the exclusion of all else, criticism becomes disconnected from its intent to analyze and critique.[9] Obviously, current social organizations and media images are not always beneficial to women.

For example, I happen to like the film *Barb Wire*, starring Pamela Anderson Lee. Just because I happen to find particular parts of the film enjoyable does not necessarily mean I think the film presents a positive image of women or of life in general. I like the fact that she wears tight black leather with stiletto heels: I think she looks pretty. However, the pleasure I take in her appearance can not be separated from the sexist culture in which I live and which forms my thoughts or from the cultural association of the black-leather-garbed dominatrix as a male lust object. For feminists, accepting media that present sexist or stereotypical views of women without criticism advances women's rights nowhere.

In *Prime-Time Feminism*, Bonnie Dow explains that the current problem facing media researchers is to find a research method that takes into account cultural context and the fact that audiences are active while still examining how the structure of film and TV contributes to shaping audience readings. Dow points out that in the United States, media like television exist first and foremost to sell products. TV series are, in fact, ads for the advertisements, and so media producers wish to appeal to as many viewers as possible while repelling no one. To do so, they adopt what Dow calls a variety of "rhetorical" strategies that can be revealed through textual analysis. Meaning, in other words, is conveyed through structure.

A simple example of how structure can influence meaning can be seen in the following example. The sentence "The cat bit the rat" conveys a sense of the cat's activity. In contrast "The rat was bitten by the cat" displaces the focus from what the cat did to what had been done to the rat. Likewise, Dow sees media forms as being like a conversation or speech that is designed to convince and persuade. For example, some political speeches are more convincing than others, and this can, in part, be traced to the speaker's choice and placement of words.

What is the prevailing rhetoric of TV? Women are the prime target audience for advertisers and TV; accordingly, TV has tended to treat women more favorably in its representations than film does.[10] However, it is crucial to note that television producers are less interested in creating shows which appeal to women than they are in building (or reinforcing) an identity for women that is favorable to what advertisers hope to sell. Advertisers resist characterizations of women that would "inhibit their incentive to purchase certain products." [11] For example, it would be rare to see a show sponsored by Revlon whose heroine did not wear make-up. Television shows, while superficially promoting images of independent women, always depict them in ways that will sell products, showing women who are well dressed, well groomed, young, and so forth.

Dow and others argue that although texts are not closed to all but one interpretation, neither are they open to *any* interpretation. Philip Green in *Cracks in the Pedestal*, for example, says that although readers from certain subcultures may appropriate texts for their own purposes, there are dominant patterns of storytelling in the media. These patterns are reflective of ideology—cultural ideals, attitudes, and beliefs that support the governing myths of our American culture.[12] Green also points out that media producers in Hollywood are people, just like the rest of us. We are all able to tap into a common culture to some extent, and the media communicate to us through this common ground of understanding.[13]

Our sense of reality, of "what is," may thus be portrayed by the media. At the same time, it can be argued that the media shape our sense of reality. Annette Kuhn has argued that the idealized images of women on TV may have a "a very real correlation" to who we think real-life women are and can be.[14] The images we see on TV are part of how we learn to see ourselves as men or women, gay or straight, and even how we see our bodies. Similarly, although Dow believes that audiences know that TV shows, for example, are fictional, "to deny that they influence our thinking about women, women's roles, and the impact of social change is, it seems to me, to be dangerously naïve. It is possible for television to be acknowledged as fiction and yet be experienced as realistic in its characterization or treatment of issues."[15] Dow's analysis of shows like *Murphy Brown*, for example, is concerned with the way that TV has helped our culture define the meaning of feminism. She is not stating that television has the power to make up anyone's mind, rather, she argues for the "hegemonic" power of television.

Hegemony is a term that describes general cultural beliefs and governing ideologies of a culture. An example of a hegemonic belief would be the idea that "in America everyone is equal if you work hard enough." This is a governing ideal that shapes our beliefs, attitudes, and actions towards others. Many people wholeheartedly believe that everyone in America is equal in spite of the fact that in the United States how "equal" you are is often determined by the color of your skin, the size of your wallet, or your gender—all of which influence how "equally" you will be treated in the workplace, by the judicial system, and in other everyday aspects of our life.

Given the fact that evidence everywhere contradicts the idea of equality for all, why is it believed by so many? Obviously, there are benefits to the belief that "everyone is equal if you work hard enough." For one thing, it reassures people that all is fair. If your next-door neighbor is a millionaire but you are living in squalor, you can feel happy knowing you are just as good as she is. She can feel happy knowing that you are poor because you are lazy or did not work hard enough; it is your "choice" to be poor.

It is important to note that the maintenance of hegemony is actually a *process*. Hegemony changes and evolves over time, at least in part because some people see through the myths. When that happens, hegemony has to adjust itself to incorporate these new values. For example, during the American Revolution, "equality" was really the equality of rich, white, male landowners. The equality of people of color, poor men, and women was never considered. Obviously in the twenty-first century, our understanding of what "equality" means has changed considerably from this original conception, thanks to women and other minorities who have demanded inclusion.[16]

The idea of hegemony as a process, as a set of beliefs that constantly changes and has to rewin us, shows that at least some people are resistant to the dominant ideal.[17] As Dow says, "the question becomes, then, not whether or not resistance or oppositional thinking exists...but how powerful or effective they are relative to hegemonic processes and, moreover, what role the media play in facilitating them."[18] In other words, precisely because resistant ways of thinking exist within individual audience members, texts must rely upon their ability to persuade their viewers.

Like Jim Collins, who believes that texts create ideological arguments through their internal structure,[19] Dow views media texts as rhetorical constructs—persuasive arguments—that make:

> [S]ome ideas, positions, and alternatives more attractive, accessible, and powerful to audiences than others....A rhetorical perspective, then, assumes that symbolic acts function to accomplish an end, and that they do so through the employment of strategies which influence audiences. However, this is not a one-way process in which texts impose meanings upon audiences. As Burke indicates, rhetorical discourse is strategic because of its ability to interact with, draw from, and respond to the context in which it occurs and in which its audiences are situated.[20]

Elizabeth Traube likewise says, "to recognize the active role of audiences in constructing textual meanings is not to ignore or deny the influences exerted by texts."[21] Traube believes that text-based studies can analyze "textual pressure or constraints on receiving audiences."[22] In other words, we still need to examine how the internal structure of a text limits the range of meanings audiences are likely to take from it.

Dow also clarifies exactly what is meant by the ability to perform a "resistant" reading. Responding to John Fiske's *Television Culture*, for instance, she observes that the possibility of audiences taking different meanings from a film does not dismiss the fact that a "dominant" ideology is usually apparent in most texts. In fact, in order for filmgoers to do a "resistant" reading, they first have to *react* to a dominant reading.[23] "In short, resistance or opposition assumes that the viewer 'gets' the preferred meaning of the text...prior to resisting it."[24] Thus, for example, it is precisely because I know that *Star Trek's* Captain Kirk is a he-man womanizer that I choose to see his character as being deeply in love with Mr. Spock. Dow argues that although "audiences *receive* essentially similar meanings from television texts [they] may *evaluate* those meanings differently."[25]

Dow argues that although it is easy to imagine all readers as resistant, it is necessary to look at audience reactions on a practical level and recognize the "differential resources and efforts necessary to construct truly oppositional or resistant meanings."[26] Preferred readings don't merely rely on the power of the text but on "competencies and codes" acquired through daily life. For example, when we watch a movie we can usually immediately distinguish between the villain and the hero. We can do this because the movie is using visual codes with which we have all grown up. The villain, for example, may wear black, while the hero wears white. Because we are all so familiar with these codes, it is very easy to make these interpretations; we don't even have to think about it.[27] In contrast, it takes more work to go against the meanings and associations we have been trained to see.

In order to do a resistant reading, an audience member also needs exposure to ideals and "ways of seeing" that are not as readily available in the American media culture.[28] For example, a person who knows little about the reality of the feminist movement might unquestioningly accept the frequent media depictions of feminists as "man hating" or "bra burning," while a person who has been exposed to feminist goals and values would laugh at the inaccuracy of such views. It is harder for the majority of people to develop familiarity with resistant ways of looking than it is to develop ease with the codes and conventions of preferred readings that surround us each and every day.

One reason for this is the repetitive and pervasive nature of television and its texts. Not only are media images all around us, all the time, but the kinds of images we see repeat themselves.[29] When we watch a cop show like *NYPD Blue*, for example, our reading of it develops in association with our memories of other cop shows, like *Hill Street Blues*, that share similar methods of presentation. The "gritty" tough guy values of *NYPD Blue* may likewise be echoed in a movie showing down the block or reiterated in the values taught by your high school basketball coach. Media texts are not isolated but are interpreted in relation to each other and the culture surrounding them."[30]

Finally, Dow argues against the view that textual analyses are only examples of how an individual researcher sees a show. She says that this view suggests that audience research is more authentic, more likely to reveal "the truth." This view actually ignores the fact that audience reception studies are done by researchers who are themselves interpreting audience interpretations. Therefore, neither audience studies nor textual-based research is inherently more authentic or more "true" than the other. Textual analysis offers a perspective different from audience reception studies but one that is equally valid. Arguing that all criticism should be "an exploration, with the unavoidable twists and turns, towards the many, sometimes contradictory, possibilities of understanding," Dow asserts that "even when a critic writes as though s/he is 'discovering' or 'revealing' a meaning in a text...what s/he has 'discovered' is the *possibility* of meaning."[31]

Dow, like Traube, thinks of audience studies and textual studies as complementary methods. As long as they are done with an eye towards cultural context, textual readings are helpful because they allow a critic to examine the dominant discourses at play in a text and connect them to the culture. Textual studies are useful because they can show how the media attempt to make sense of social change, like the changes brought on by the feminist movement, which, in turn, might influence how audiences come to think about a topic.[32] Neither Traube nor Dow consider textual analyses as providing a pure reflection of reality. Rather, they seek to reveal how media work to limit and define areas of public discussion.

It is important to note that Dow's focus on a textually informed understanding of public discourse is not about determining the one "true" meaning of a text. Nor is it a return to the "images of women" school. In any case, as Traube comments, the problem with the "images of women" approach was not that it tried to connect film to wider social currents but that it was an oversimplified approach, which saw film "as a reflection rather than a *mediation* of its social context."[33] Mediation implies a give-and-take relationship between text and context.

As Traube explains it, Christine Gledhill has previously argued that theorists need to take into account diverse social audiences and differences such as race and class as well as gender. We also need to examine how media interact with power relations in real life. Media studies should thus focus on the ways that media texts and audiences are engaged in a process of negotiation, focusing on "the multiple and contending possibilities of meaning embedded in a text as these might appear to particular social subjects."[34]

Following this framework, Dow interprets the ways sitcoms have constructed particular attitudes towards feminism. She describes a set of dominant discourse practices that she believes were shared by both television and "secondary texts" such as magazines, news stories, advertisements, etc. Following Todd Gitlin (*The Whole World Is Watching*) and Lynn Spigel and Denise Mann (*Private Screenings*), Dow focuses her analysis on the ways sitcoms like *Mary Tyler Moore* and *Murphy Brown* have functioned as part of the cultural debates about feminism and how TV programming has contributed to the "cultural conversations about feminism in this country at different historical moments."[35] In Dow's view, TV negotiates social issues, offering various interpretations of their meaning. Because there is a range of meanings rather than one single meaning, more people will find something with which they can identify or enjoy in the show. Along with secondary texts like ads or magazines, television describes "the main lines of the cultural consensus about the nature of reality."[36] Rather then offering one single way to think about feminism, TV offered several, sometimes contradictory, ways of thinking about this social issue.

By examining both the historical setting and other media during the time the programs first aired, Dow shows how TV helps produce, transmit, and transform cultural ideas about what it "means to be a 'liberated woman.'"[37] It is important to reiterate that Dow does not find television representative of "the way it is." Television meanings are often contradictory. In fact, the persuasive function of TV is not to "provide solutions to cultural conflicts but, rather, to negotiate their parameters for debate."[38] TV doesn't necessarily tell you *what* to think, but it can influence *how* to think: It sets up some issues as relevant, while others are made invisible through nonrepresentation, for example. It also constructs a certain set of ways to consider a topic but excludes others.

Dow looks at how textual components such as genre, character development, and story lines react with the kinds of cultural discourses about feminism that were occurring at the time the shows first aired. She examines how feminist rhetoric and events were "absorbed, structured, and represented in media discourse for popular consumption."[39] Dow's approach is similar to that of Linda Williams' analysis of *Mildred Pierce,* in which she shows how the film deals with tensions in gender and class that were specific to the period in which the film was released.[40] In *All That Hollywood Allows*, Jackie Byars similarly contextualizes her studies of 1950s melodramas by situating the contradictions she finds in the text against an understanding of the 1950s as a period of similar tension regarding the role of women in society. Similarly, Julie D'Acci's case study of *Cagney and Lacey* illustrates "how during a particular and important moment woman, women, and femininity were defined and constructed in the cultural sphere of fictional representations and interpretations."[41]

The idea of textual analysis done with an eye towards social context is important. Viewing texts as rhetorical allows us to examine texts for the means by which they channel viewers into certain ways of seeing. Although individual viewers can choose to evaluate the messages of the preferred reading negatively or positively, depending on their personal preference, each viewer nonetheless "gets" the point. This approach is free from the vulgar reflectionism of the images of women school. Because it leads to an understanding of the kinds of public debates circulating in the culture at large, it allows us to focus on the process of how texts *negotiate* meaning.

Thus, textual and audience studies may be viewed as complementary approaches. The use of one method depends on the type of questions one is asking. If one is concerned solely with the effects of the text on the viewer, audience studies may the best approach. But since texts are not open to every interpretation, textual studies are necessary in order to shift between an understanding of the text and the audience in order to gain an understanding of what possible meanings the texts are capable of forming for different audiences.[42]

CHAPTER TWO

Interpreting Heroic Action

Female Action Heroes

Analyzing how meaning is *negotiated* within a text, as well as outside of it, is particularly useful in regard to examining contemporary images of women and feminism, which tend to be highly contradictory. Since the seventies, ideas of what it means to be "a woman" have changed drastically as have images of feminism. It is important that there is not one strand of discourse rather, there are many competing discourses about feminism and the status of women in our society.

For example, it is currently common to hear that feminism is dead: either because women are now equal and feminism is unnecessary or because feminism created more problems than it solved. Equally present is the idea that feminism is fractured. Feminists are not a unified group of women; instead, various kinds of feminisms exist for different groups, and these various groups do not necessarily support each other. The idea of a conservative backlash against the feminist movement and women is also in existence. There are also a wide range of images of women: Women are "free;" they have "girl power"; women are divided by race against each other; women are victims; there is a backlash against women.[1] All of these discourses about feminism and women are circulating in the early twenty-first century. Against this context, supporting it, reacting to it, and perhaps changing discourses in turn are contemporary images of women in the media.

One particular notion that has emerged in public conversations over feminism and the state of women in contemporary society is the question of female power. What is it? Where is it? What does it mean? The female hero is a great place to investigate the meanings of female power circulating in society because she so visibly assumes a role defined as masculine and powerful. As the "star" of the series the female hero not only assumes the central role but destroys conventional ideas of the female body as passive, as to be looked at, as controlled by men. The female hero *takes up space*.

The female hero has appeared on television since the sixties with such characters as Honey West and Emma Peel. However, she is a very interesting figure to examine at this particular time because her recent emergence into the foreground of film and television programming seems to echo the growth of female physicality in other arenas. For example, the growing fitness movement in the United States superficially suggests an increasing comfort level, at least among women, with female physicality and power. Female athletes are gaining more attention and popular support than ever before as in the recent inception of the WNBA. And advertisements, like those by Nike, hint that physical strength for women is now acceptable.[2]

Despite this seeming acceptance of female physical power, however, researchers in the area of women and sports have found that the growing number of images of athletic female bodies does not necessarily reflect the acceptance of feminist values such as female empowerment. On the one hand, images of active women may empower individual women, but on the other hand, these images are not necessarily progressive from a feminist perspective.

For example, Mike Messner shows how female body building, which on the surface goes against traditional images of femininity, echoes many of the same themes seen in the beauty culture with an emphasis on self-judging and diet products. Moreover, Messner shows that since body building revolves around "macho" aesthetics like the building of muscle, the "fit" body is still seen as male—strength and male abilities are still valued over the capabilities of a female body. Similarly, in her book *Unbearable Weight*, Susan Bordo elegantly reveals how the fitness industry uses the language of female liberation and female power by stressing choice and strength to promote women's adherence to even more rigid standards of bodily maintenance. These standards—such as working out three hours a day in the gym to achieve a "toned" look—actually constrain and disempower women.[3] Focusing on maintaining a "tight" body keeps women from other activities in the public realm, like changing the world.[4]

Contradictions are apparent in representations of other kinds of "aggressive" women as well. Kathleen Rowe has written about the "Unruly Woman," a stock character that appears in various forms throughout popular culture. The Unruly Woman is a figure who turns traditional images of femininity upside down, most noticeably through "unfeminine" behavior. For example, the Unruly Woman takes up space through her large size, disruptive behavior like spitting, belching, or farting, and outspoken humor. Rowe shows that while women are usually made the object or butt of jokes, the Unruly Woman is the joke-teller.[5] By telling jokes, she places herself center stage and claims the right to public space—a space that has historically belonged to men. Patricia Mellencamp has likewise pointed out the "unruliness" of comedians like Lucille Ball. Such comedians are transgressive partly because, like the female action hero, they take up space with the physical movement of their bodies at the same time that they disrupt life in the public sphere.[6]

The Unruly Woman may be a rebellious figure in popular culture, but she is also frequently hated. Rowe points to the comedian Roseanne's treatment by the press to demonstrate how the Unruly Woman is reviled and disparaged at the same time that she projects a vision of female power.[7] In addition, Mellencamp writes that while texts like *I Love Lucy* offer moments of disruption, the Unruly Woman is usually constrained by narrative closure. For example, the rebellion implied by Lucy's physical antics is ultimately denied: Lucy is scolded, chastened, and brought home by Ricky at the end of each episode.

It is clear from these examples that the meaning of female activity in contemporary media representations is unclear. It would be inaccurate to see the role reversal of physically active women simply as a sign of feminist progress. It is necessary to closely study such images in order to discover the full range of potential meanings in their representations of active women.

Like images of other unruly women, images of female heroes are frequently contradictory. Shows from the seventies like *Police Woman*, *Get Christie Love!*, *Charlie's Angels*, and *Wonder Woman* departed from existing television representations of virginal white womanhood. They introduced, as Julie D'Acci has observed, "sexual dimensions to the traditionally more domesticated woman, revealing her as a sex and beauty object."[8] At the same time such shows emphasized the sexuality of their stars, which was dubious progress, they did break barriers for women, placing both black and white females in leading roles. Thus, these shows, part of a "jiggle" phenomenon, demonstrate how TV works to capture various segments of an audience. They simultaneously appealed to proponents of the women's movement and served as eye candy.[9]

In *Cracks in the Pedestal*, Philip Green likewise concentrates on the ideological implications of female-centered narrative as a means to examine how Hollywood has responded to the feminist revolution. Green argues that the presence of female heroes are "tokens of our fragmented reality," a sign of "the divided consciousness of a hegemonic ideological structure that is under severe attack."[10] In other words, the changes wrought in dominant culture by the women's movement has resulted in the fracturing of that dominant culture, opening up space for oppositional images to emerge at least within marginal forms of entertainment.

Green argues that within certain subgenres of action adventure films, particularly martial arts films, female action heroes like Cynthia Rothrock have achieved, within the structure of the text, a kind of equality with their male counterparts in that it is possible to imagine the female hero, "if only in fantasy," as "the individual" and "the mythic bedrock of liberal capitalism."[11] In his comparison of Chuck Norris's film *The Octagon*, with Cynthia Rothrock's *China O'Brien*, for example, he notes that both films are about vigilante justice and recuperate the "overall outline of dominant ideology" of American individualism.[12] Similarly, he believes some texts like *Xena* feature a point of view that is clearly female.[13]

This equality is less apparent in mainstream Hollywood. Moreover, it is important to recognize that Green does not find a mere role reversal occurring. The gender of the "individual" does influence the reading of the film. While Norris's vigilante projects a quality of sadism, O'Brien is a hero partly because she is able to do what no one expects her to do, and she does it on behalf of her community.

Thus, as Green observes,

> [T]he standard (male) version recuperates it as the strong protecting the weak against the ignoble forces of due process: as incipient fascism. The female version emphasizes the genuine conflict in the dominant ideology, the tension between community and individuality, without allowing them to be welded together in the patriarchal form of authoritarian populism: the woman as hero cannot do that even if she wanted to.[14]

Green's analysis here stems from his understanding of the ways women in action texts occupy the position of underdog, which is a central figure in the tradition of American individualism. However, what Green does not detail, but what I find crucial to consider, are the precise ways that her gender produces the sense that the female hero is the underdog. What are the textual strategies employed that frame the female hero in ways different from the male hero?

Green's ideological project means that he is mainly concerned with narrative structure rather than other elements of signification that portray the female hero quite differently than if she were Jean Claude Van Damme. For example, despite functioning as the "individual," she is often surrounded by a bevy of friends, who support her, fight for/with her. Instead of having a "babe" to rescue, the romantic interest usually fights better than she does. In *China O'Brien*, for example, Rothrock's ability to fight is explained away, as it is in many action films, by her alignment with her dead father and brother. As Yvonne Tasker has argued, these clues signify the female hero's identification with the male, her separation from regular women, as well as creating the sense that she is an unformed adolescent in search of identity.[15]

In addition, Rothrock is more likely to be physically hurt and rendered incapable of fighting. We can hear her grunting, breathing heavily, crying out in pain, all of which tell us that fighting is *hard*, and she is about to be overwhelmed. This may provide a laudable lesson in cooperation and the need of support from a community, but, within the individualistic framework of action films, it also presents an image of a woman who, however physically fit, is still not on par with the male and needs to be protected by her (male) friends.

A narratively inspired ideological analysis does not tell us how the gender of *China O'Brien* creates a different reading from that created by a Jackie Chan film, whose films frequently have an emphasis on community that, from an ideological standpoint, is often similar to the one Green discovered in the *China O'Brien* films. Chan, particularly as he moved away from Bruce Lee-inspired films to his own particular brand of Buster Keatonesque martial arts, similarly tends to show the effects of violence, an awareness of vulnerability, and an idea that it is at least partly his intelligence and help he gets from his friends, and not just his fighting skills, that help him win. As Tasker has noted, at times "he seems to win his fights more by good luck and determination than skill."[16]

Tasker's own textual reading of Chan and his "softness" is offset by a closer reading. Extratextual knowledge gleaned from secondary sources and other Chan films play a huge role in the meaning of Chan's films. There is widespread knowledge that Chan is an *extremely* skilled fighter who regularly risks death performing his own stunts. Hapless he might *appear* to be, but it would be hard to find an audience member who does not see Chan's apparent helplessness as part of his skills as a trickster. One waits confidently for the moment when his true fighting ability emerges.

Focus on narrative alone and the isolated text misses such extratextual knowledge. It also may ignore crucial questions of gender and race. Chan, a man, plays a similar sort of hero to Cynthia Rothrock's *China O'Brien*. How are North American readings, in turn, affected by the fact of Chan's race, considering that Asians in Hollywood films have historically been presented as weak and feminine compared to the white male? Chan might be so successful in the West because his films structure him as underdog, thereby removing the threat of his non-white masculinity. Chan's ability to play "soft," as Tasker calls it, thus grants him greater acceptance among western audiences than any previous Hong Kong action star, yet this acceptance allows him to eventually emerge as "tough." What does it mean that Cynthia Rothrock, a woman, already "soft" in our eyes, plays a "tough" action hero but plays tough "softly"?

The point is that simple narrative analysis does not reveal any of these complexities. It is perhaps not surprising that Green concludes that the female action hero remains trapped by Mulvey's postulation of the female as object:

> Their initial gender inversion aside, hardly any of these films deviate from the conventional visual, moral, sexual (including racial) codes of dominant culture....Above all, Mulvey is certainly correct that, given the extent to which a particular relationship among gender, power, and the male gaze is inscribed in contemporary culture, any visual presentation of women (or girls) as spectacle, especially in the context of violent action and its directly visceral appeal, is almost certain to emphasize the fetishistic structure of our (male) viewing....Thus for most of the movies in the female action/adventure genre, the camera's gaze remains male.[17]

While he states that it is impossible to read figures like Xena and Ripley as *just* existing for the sake of male pleasure, Green finally sees the spectacle of the female body as pinned by the male gaze. However, there is an extent to which the camera can *invite* a fetishistic gaze. And that is something that can be controlled by film techniques. What happens when viewers are able to access texts whose camera-work doesn't explicitly fetishize female bodies? Conditioned as they are, many viewers may choose to fetishize the body, but are there texts which present the *possibility* of a different way of viewing?

It is also important to ask whether Mulvey's ideas may be as usefully applied to television. Mulvey's analysis is based on classic cinema whose filmic conventions often differ from television. Television sitcoms, for example, often privilege long shots over close-ups, which certainly changes the ways the body can be fetishized by the camera. Similarly, as Kathleen Rowe has observed, Mulvey's ideas fit less well with comedy and action genres. Although the position of woman-as-object remains in many action/adventure narratives, the film conventions of action movies allow greater freedom for the female action hero. As do many comedies, action films tend to show the entire bodies of female heroes, showing them as figures in motion against a landscape. Their bodies thus seem part of the world around them rather than set apart. This contrasts to the classic films Mulvey looked at in which the female body was often filmed in close-up, fragmented, giving the sense that the woman is set apart from the world around her.

However, even if we say that female heroes remain a visual spectacle, this is not necessarily a bad thing. Indeed, as Rowe has written, spectacle is a way that women can claim visibility in the public sphere. This sense of spectacle

> differs from Laura Mulvey's....More Foucauldian than Freudian, it suggests that visual power flows in multiple directions and that the position of spectacle isn't en-

tirely one of weakness. Because public power is predicated largely on visibility, men have traditionally understood the need to secure their power not only by looking—but by being seen—or rather, by fashioning, as author, a spectacle of themselves. Already bound in a web of visual power, women might begin to renegotiate its terms...by utilizing the power already invested in us as image, we might begin to negate our own 'invisibility' in the public sphere.[18]

In other words, it is not the to-be-looked-at-ness of women that is the primary problem but rather the interpretation of that quality. Women seen as spectacle may be read as being in power as long as they place themselves as author, in control of their own image. Similarly, the spectacle female action heroes provide to others is less important than whether the spectacle encourages the viewer to read the female hero as lust object or as subject of her own actions. It is the small, subtle signs that may answer these questions, and thus it is imperative to do a close, textual reading of action texts.

As Yvonne Tasker has noted in *Spectacular Bodies*, it is necessary, especially in action cinema, whose "narrative" is in a large part the visual image, to take into account the "tone" of a text. The tone is "indicative of a play of elements which cannot be understood through an ideological analysis based on narrative structure alone."[19] In *Working Girls*, she examines how films such as *Terminator* work to "reinscribe, in different ways, the *female* body in terms of masculinity," noting that the same metaphors of softness and hardness in play in male-centered texts are at work in action films starring women.[20] She coins the term *musculinity* as indicative of the "extent to which a physical definition of masculinity in terms of developed musculature is not limited to the male body."[21]

As she says,

muscular cinema, rather than signaling a radical break with the past, inflects and redefines already existing cinematic and cultural discourses of race, class and sexuality...it is worth attempting to a give an account...[of] the complex ways in which popular cinema affirms gendered identities at the same time as it mobilizes identification and desires which undermine the stability of such categories.[22]

For Tasker, the emergence of muscular female action heroes in movies not only signals a response to feminism but, more importantly, poses problems for binary concepts of gendered identity. If hardness is male, then how do we read a hard female body? She thus asserts that it is not possible, when studying action films, to rely upon Mulvey's binary conception of active/male female/passive, since both men and women are positioned as spectacle at the same time they become more than spectacle.[23]

However, Tasker acknowledges that many images of the female body in action, whether they are overtly sexualized or not, bring to mind the images of "female independence derived from or developed within sources including advertising and soft pornography."[24] In other words, it is hard to look at a female hero without instantly thinking of the dominatrix, for example. Nonetheless, these associations do not "fix the meaning" of the active female body since there is "a tension between an image built, designed for contemplation in static poses, and the situation of such images within the context of action."[25] In other words, *movement* and its qualities are essential to address when discussing spectacle in action films.

Tasker has provided some framework for understanding various representations of active women in film. In *Spectacular Bodies* and *Working Girls,* she categorizes different types of female action heroes and also examines the interaction between image and narrative. She concludes that many movies offer an image of independent women who break from the white, middle-class view of a passive and hysterical "femininity."[26] At the same time films attempt to explain away the problem presented by the female hero's active body:

> The female action hero offers a fantasy image of (proletarian) physical strength showcased within narratives that repeatedly seek to explain her (and to explain her away). Female action heroes are constructed in narrative terms as macho/masculine, as mothers or as Others: sometimes even as all three at different points within the narrative.[27]

Thus, the aggressiveness of the "maternal" female action hero is explained by her desire to protect her young. Maternal motivating factors occur in such films as *Aliens*, *Terminator 2*, or *Strange Days*. The figure of the tomboy, seen in such films as *China O'Brien* or *Twister* is apparent in films in which the female hero is identified with her father and is consequently "in search of authority, and sometimes reconciliation with authority."[28] Importantly, these tomboy figures are often asexual at the film's beginning. Thus the tomboy's desire to act aggressively is read as a stage of adolescence on route to accepting "true" womanhood.[29] As in romance novels, the narrative is a journey of discovery in which feminine sexuality is ultimately recuperated into the patriarchal order.[30] The third figuration that Tasker has observed is the "fetishistic figure of fantasy" derived from comic books and apparent in films like *Barb Wire*, a figure that links female aggressiveness to an exaggerated female sexuality.

Moving between these three modes is the feisty female hero, simultaneously appearing as a romantic interest or sexual object as well as a protagonist. Importantly, her feistiness is acquired through association with men—through her identification as the hero's girlfriend, for example. In addition, within the discourse of the movie, the feisty female hero is described as adult, heterosexual, and as clearly different and set apart from other women.[31]

Women also act as sidekicks to men. These sidekick roles are related to but different from buddy partnerships. While buddy movies tend to have a level of equality between the hero and buddy, the sidekick takes a decidedly inferior position in the hierarchy of knowledge and skill, as, for example, in *Eraser*, in which Arnold Schwarzenegger plays an agent protecting a witness, Vanessa Williams. Though Williams does fight in the film, she is clearly presented as an amateur, whose inexperience is underscored by Schwarzenegger's status as a professional.

Sherrie Inness's book *Tough Girls* examines elements such as narrative, costuming, and characterization to study the changing nature of female heroes in film, comics, and television from the sixties through the nineties. Her chapters on recent films like the *Alien* series and *Silence of the Lambs* and television series like *The X-files* and *Xena: Warrior Princess* are the most useful to examine here. She concludes that, while undercut in particular ways, characters like *The X-files'* Scully and Xena do promote an image of the "tough girl" as self-sufficient and powerful.

Despite her voluptuous costumes, Inness observes that "[i]t would be a mistake to view Xena as primarily a sex symbol," noting that she is "tougher than virtually all women in similar action adventure shows."[32] Inness points to such qualities as Xena's independence from male characters, her self-reliance, her use of her body as her main fighting tool rather than magical implements like those used by Wonder Woman, her lack of concern over how she appears to men, her relationship to her sidekick, Gabrielle, and the series' relatively ambiguous depiction of her sexuality. Nonetheless, Inness finds that although *Xena* alters some traditional conceptions of the female hero, it reinforces others. For example, Xena is white, beautiful, and highly desirable. Moreover, the prominent use of camp and humor diminishes her threat.[33] Yet Inness finally concludes that despite the contested nature of her depiction, Xena ultimately provides a "hero who makes viewers recognize that women can be just as tough as the boys, while questioning what it means to be tough."[34]

The studies discussed above clearly demonstrate that women in media do not occupy strictly passive roles. They help to articulate the ways that media representations of women are changing and are suggestive of cultural changes in regard to women's public power. They provide useful models for helping us understanding the social meaning of the female action hero. Above all they point to the ways that there is no singular "meaning" of the female action hero. She is frequently a contradictory figure, functioning simultaneously as eye candy and as a figure of power, reaffirming the notion of strength as a "masculine" quality at the same time that she reveals the falsity of this assumption.

Warrior Women takes a slightly different approach to the topic of female heroes than these previous studies. I am less concerned with determining the explicit political ideology of the text or outlining the various "masculine" or "feminine" qualities of the hero or her relative level of "empowerment" than in how female-centered action texts reflect particular attitudes towards the body. As outlined in the introduction, the body is not only a material object or a particular performance of social identity but a process, "a way of knowing and marking the world, as well as a way of knowing and marking a 'self.'"[35]

As will be discussed in the next section, male-centered action narratives feature specific imagery that conveys particular cultural attitudes towards the body. These attitudes influence our understanding of identity, what it means to be a "hero" as well as our behaviors and attitudes. *Warrior Women* examines the central images, metaphors, and frameworks of female-centered texts in comparison to those of male-centered texts. What is the imagery at play in female-centered texts? What does this imagery convey to us about how bodies are valued, the "proper" use and behavior of the body, the relationship of the body to the external world, and the relationship of the body to heroism and female power?

Male Action Heroes

One way to begin examining the meaning of the female hero is to contrast their representation to that of male heroes. Jane Tompkins's analysis of Westerns is useful for this discussion because it describes bodily images that are common to not only the Western genre but to the action genre as a whole. According to Tompkins, Westerns function as a symbol of freedom, providing opportunities for conquest as well as an escape from society.[36] They reveal a desire for a self-transformation with life on the frontier being "a way of imagining the self in a boundary situation—a place that will put you to some kind of ultimate test."[37] Like all action texts, Westerns ultimately function as a search for meaning told through the body of the hero.[38]

Tompkins traces the development of the Western to a cultural background in which the cult of domesticity held sway over the American imagination. She observes that nineteenth-century sentimental novels popularized middle-class values of the times, with their sentimental focus on women, the private spaces of the home and self, and spirituality. At the same time, the United States was experiencing an invasion of women into public space.

Tompkins argues forcefully that the Western must be read as arising in response to this situation and thus as expressing the fear men of the time had of "losing their mastery and hence their identity."[39] Westerns turn on "external conflicts in which men prove their courage to themselves and to the world by facing their own annihilation."[40] Tellingly, within the fictional space of Westerns, women were excluded from the narrative or objectified; the emphasis in Westerns is on the "importance of manhood as an ideal."[41] This definition of manhood is based on a fantasy of mastery and dominance over the physical, material realm. Heroes dominate other men, women, and animals and their own bodies.

The typical hero is enigmatic and silent. Because speech suggests incompleteness and a need to be in relation to another, silence is a way of keeping the body impenetrable, of policing physical boundaries.[42] In distrusting language, Westerns create a sense that only objects—concrete *things*—are real compared to nebulous emotions, sensations, and human relations, those qualities we associated with the interior body and the "feminine." Tompkins also finds that the heroes' rejection of language coincides with their drive to merge with the dangerous terrain. Language implies incompleteness: you talk to or with another. It also implies self-consciousness. In contrast, heroes seek to lose all sense of self-consciousness. They aspire to *be* the perfect landscape, in other words, to become perfect *in themselves* rather than a self-in-relation.[43]

Towns, representing social order, are paradoxical in their function. On the one hand, they are a haven, a restful place. One the other hand, they threaten to "entrap the hero with intimacy, mutual dependence and a network of social/emotional responsibilities."[44] Towns, like language, threaten the bodily integrity of the hero. Rugged landscapes, in contrast, help define the man. Westerns abound with descriptions of the body's sensations in the environment. The hero demonstrates his mastery, the control of his mind and spirit over his body, by surviving in a dangerous desert environment. The landscape can thus be understood as a metaphor for the body. Similarly, the heroes often master powerful horses that become their partners. The recurring motif of suffering horses is analogous to the idea that "will is to body as man is to horse."[45] Just as the heroes must suffer silently, the suffering of horses reveals the "aim to dominate emotions and control the fleshy mortal part of the self and the external material world."[46] Through such images, Westerns teach the ideology of domination and control of others as well as the self.[47]

In addition to defining masculinity, Westerns also focus on death. Tompkins believes this represents a forcible rejection of both the "feminized" Christian beliefs of the middle class as well as the eternal spirit.[48] As she notes "once you no longer believe you are eternal spirit, risking your life becomes the supreme form of heroism, the bravest thing a person can do."[49] In losing the spirit, the body becomes the locus of battle. Paradoxically, however, the struggle of the body is the means through which transcendence is achieved.

Tompkins observes several similarities between the Western hero and ascetic monks. She writes "[t]he monolithic, awe-inspiring character of the landscape seems to reflect a desire for self-transcendence, an urge to join the self to something greater."[50] Similarly, the bleak existence in the desert is the place where

> language fails and rocks assert themselves. But by the same token it is the place where something else becomes visible, an ineffable thing that can not be named. The desert pushes the consciousness of the hero and of the reader/viewer beyond itself and into another realm....It is not only the body that is tested here; the desert is a spiritual proving ground as well.[51]

In other words, heroic narratives are about the journey of an individual to a transcendent state.

Tompkins's discussion of the Western is useful in providing a foundation with which to evaluate contemporary female-centered action texts, both for their similarities and differences. Westerns promote a distinct way of viewing the world that is based on mastery of the self and others, a philosophy that uses the punishment and denial of the body as a means to spiritual transcendence. Of particular importance in discussing representations of the heroic body are Western motifs such as the use of boundary situations, the relationship of the hero to landscapes and environments, the construction and definition of "heroic" identity, the tension between community and individual, and the search for some sort of ultimate "meaning" vis-à-vis death, enacted through the drama of the protagonist's body.

It is crucial to recognize that these motifs are not unique to Westerns but appear in various forms throughout action genres. In almost every contemporary male action film, one can trace a similar path in which the transcendence of the hero ultimately depends upon the rejection and punishment of the body and all things associated with it, such as the feminine. Susan Jeffords, for example, discusses the ways the "hard bodies" of male action heroes of the eighties in films like *Rambo, Lethal Weapon, Robocop*, *Terminator*, and *Die Hard* allowed viewers to see themselves as masterful, in control of the environment, and dominating those around them. Like the bodies of Western heroes, these bodies reinforced the need to maintain strong boundaries between self and other.[52] Thus, the hero of *Lethal Weapon* protects the family of his partner but "cannot afford to be burdened by the risk of a family" and the need for others that implies.[53]

Yvonne Tasker observes that, seemingly paradoxically, the action hero is frequently rendered passive, "subject to a range of restraints and oppressive forces."[54] However, she notes that this suffering "lends him a certain tragic status, and demonstrates his remarkable ability to endure."[55] Likewise, while films like *Rambo* offer moments in which the hero's body is made vulnerable, for example, placed on visual display or shown being penetrated and wounded, Jeffords notes that such moments only work to underscore the fact that the hero is unlike other men. Rambo, for example, pushes a piece of wood through his body, demonstrating, that unlike the "average" viewer, he is able to endure remarkable levels of pain.[56] Once again, his heroic quality is proven by his ability to master his suffering. He is a rugged individual. As William Warner writes, at the end of the film the audience is left with "a freeze frame image of Rambo as a nuclear subject, a self-etched against a landscape where no supporting social network seems necessary."[57]

Jeffords observes that male heroes of the nineties superficially seem to break with the tradition of separation from others, in that these films frequently feature a narrative resolution in which the hero is reconnected to a family or his softer emotions. However, in her analysis of films like *Terminator II* and *Robocop II*, Jeffords writes that "the power of individual decisiveness and individual action is drawn as paramount in these films."[58] Such films are "not a simple negation but rather a rewriting, a repetition, a retelling of the story of masculinity...though that rewriting seems on the surface to be a rejection of so many of the spectacular identifications of masculinity of the 1980's—technology, violence, power, command, strength—its mainframe is still very similar: the reproduction of masculine authority...through the affirmation of individualism."[59] The "hardness" and impenetrability of the heroic body is likewise maintained, as is seen, for example, in contrast between the "hard" but heroic Terminator with the liquid, "female" body of the villainous T1000 in *Terminator II.*[60]

While within dominant Hollywood cinema most heroes have been white, certain subgenres, such as martial arts films, have presented men of color in leading roles. Martial arts films, like other action texts, "pose questions of gendered identity in terms of visual and verbal metaphors of hardness and softness."[61] These metaphors are at play in both the textual and extratextual discourse surrounding action heroes like Bruce Lee, for example, who struggles to leave behind softness and become a perfectly hard body. The emphasis on bodily hardness once again reiterates a distinction between vulnerability and impenetrability, between "feminine" and "masculine" qualities.

It is important to acknowledge that the meaning of bodily hardness and physical endurance certainly changes depending on the race of the hero. Lee's attempt to distance himself from feminine "softness" is also an attempt to break the stereotype of the feminized Asian man. Likewise, Yvonne Tasker observes that in martial arts films the body and its ability to endure convey the hope of achieving success through the body.[62] Critics have thus linked the popularity of martial arts films with urban black audiences to a "ghetto myth" in which dispossessed groups supposedly identify with the "hero's struggle to overcome."[63] In contrast to the reality experienced by many men, the spectacular bodies of action heroes function as "sites of pleasure," because they offer a sense of identity that is strong and free of external constraints—a body that can rebel against oppressive authority. Such bodies, therefore, offer a fantasy in which the existing social order is overturned.[64]

While it is important to consider the ways the recent emergence of stars like Wesley Snipes and Jet Li affect the traditional representations of action hero, it is notable that, within mainstream cinema, until very recently it is primarily white male bodies that are "hard" enough to serve as hero. The fact that the hero's body is so often raced as white as well as gendered male conveys specific notions about the hero's dominance and his relationship to his environment and others. For example, Western heroes and the heroes of many science fiction texts share a similar relationship to their physical environment. The oft-repeated story of Gene Roddenberry peddling *Star Trek* as a "*Wagon Train* in space" is symptomatic of the assimilation of the frontier myth into action narratives.[65] A wide range of science fiction action texts, for example, explicitly posit outer space as an unknown, feminized territory destined to be discovered and exploited by the white male pioneer (hero) while its feminized "Other" inhabitants exist to be known, conquered, and dominated. In addition, as Barbara Creed has observed, films like *Aliens* or *Species* rely on the imagery of the female reproductive body to signify the terrifying, the unknown or monstrous.[66] The goal of the hero in these films is to destroy the female monster/alien and punish her for the threat she poses to patriarchy.[67]

This framework is apparent in recent action films like *The Matrix* which features a hero whose body is penetrated by technology, tied to a machine. Despite the apparent vulnerability of his penetrated body, however, the film, like much cyberpunk, reiterates "mythologies of the invading male, entering and conquering a virgin territory" and by doing so follows traditional subject/object relations between male/female, mind/body, and white/black.[68] Neo, the hard-bodied white male, enters the feminized (and monstrous space) of the Matrix. It is he, not his black male or white female helpers, who has the power to bend the realm of the Matrix to his will. His will (mind) has the power to shape the physical environment (the body) according to his every whim. Within the space of the Matrix, his omniscience is reflected by his physical superpowers.

Once again, Neo's ultimate transcendence as a hero is dependent upon his rejection of connection, particularly the "feminine" and the mastery of the mind over the material realm. The machine that runs the Matrix is depicted as a sort of monstrous mother to whom Neo is tied through the umbilical cords of wires. In a process that is clearly meant to reflect birth, Neo is shunted from a womb-like chamber, down a tunnel, covered in fluid. It is implied that he is unable to see reality at the start of the film because he still exists in an infantilized relationship with the mother/machine. In other words, Neo achieves his power only after separation from the feminine. His ability to master his physical realm is linked to his recognition of the truth of reality and his separation from the monstrous machine mother. The mind is dominant over the body.

As we see, the bodies of action heroes thus represent particular ideas about how the world functions. Heroic identity tends to be founded on the isolation of the self from others, in which the body struggles to become impenetrable and self-contained. The hero is both defined by and controls his physical environment. The "hardness" of the hero's body reflects his mastery of himself and others. Though told through the medium of the body, male-centered texts repeatedly establish the primacy of the will and mind, consequently offering a rejection of the realm of the flesh at the same time they punish or repress other aspects of the feminine. The heroic identity is thus one that that is formed through the exclusion and punishment of the "feminine"—both those qualities of the hero's self deemed as "feminine" and those of real women. It, therefore, becomes particularly interesting to examine how female heroes survive and even thrive in a genre that has been built around their exclusion.

The bodily images of silence, hardness, and impenetrability, the body's relationship to the environment and others, and the power of the mind over the body seen in male-centered texts all express particular notions about the meaning of heroic identity, mapping out the behavior necessary in order to reach heroic transcendence. *Warrior Women* focuses on the changes that occur in this imagery within female-centered texts. In particular, the analysis of *La Femme Nikita, Aeon Flux*, and *Buffy the Vampire Slayer* focuses on boundary issues such as the penetrability of the body, the relationship of the self to community or state, the relationship of the female protagonist to her physical space, the role of suffering, and the relative value the texts place on mind and body.

CHAPTER THREE

La Femme Nikita, Romance, and the Plastic Body

La Femme Nikita premiered on the USA cable network in the spring of 1997. The show is based on the French film, *La Femme Nikita* (1991), which was remade into an unsuccessful American film, *Point of No Return* (1993). Though filmed in Canada and featuring many Canadian actors and crew, the series' executive producer is the American Joel Surnow. In the original film, the heroine is a cop-killing drug addict sentenced to die. She is rescued and trained as an assassin by a secret government agency, from which she eventually defects. The television version focuses on Nikita, a twenty-something female street punk in an unnamed city, who is falsely accused of murder and sentenced to death. Before she can be executed, she is removed from prison to a secret facility called "Section One," where she is trained as a spy. Section, an anti-terrorist group whose ends are described as "just" but uses "ruthless" means, may or may not be a government organization. Her only possible ally in the agency is the ambiguous Michael, who trained her and is her immediate superior. Nikita must perform her duties while dealing with her unwillingness to become the ruthless killing machine she is intended to be.

Nikita is an interesting case to examine for both its similarities and differences to male-centered texts. As previously discussed, male-centered texts typically reflect a concern with maintaining boundaries between self and other, repressing the body, and dominating others. For example, the silence of the hero is a bodily sign of his individuality and his physical impenetrability. Although the hero may work to maintain the community or protect a family, like Shane, his separation from others is maintained. He does not belong to the public body; he is not a self in relation, constructed through interactions with others, but is a discreet entity; his body consequently suggests solidity. The landscape in which the hero finds himself frequently tests the limits of his physical endurance; he must conquer or survive this threatening space to achieve his goal. The environment thus functions as a metaphor for the body, against which the hero must struggle in order to prove his spiritual worth. By dominating his body/his environment, the hero projects a sense of control over the physical world.

Like many traditional representations of the female hero, *Nikita* frequently tries to "explain" away Nikita's presence as hero by overtly sexualizing her body, presenting her as a victim, and framing her violence as protective and maternal. In contrast to male-centered texts, however, the boundaries between self and other/external environment are more fluid: Nikita is as much affected by her world as it is affected by her. The environment of Section in which she finds herself is presented as dangerous to Nikita's integrity as an individual. The text is very critical of the way its protagonist's self (both body and mind) is controlled, penetrated, and invaded by external forces. While the series promotes an ideology of individualism, Nikita's battle with her environment is depicted as a process of navigation rather than domination. Nikita survives by adapting, learning to assume different roles within the organization. She is thus shown being formed through her relations to others and constructed by her position within the social order of Section. In addition, the repression of the body found in male-centered texts is not in play. If anything, the series fetishizes the realm of the flesh, of emotions, physical sensations, and relations with others. *Nikita* utilizes narrative and filmic techniques drawn from genres like romance and soap operas to suggest a heroic transcendence based on an acceptance, rather than a denial, of the body.

Explaining the Female Hero: Sexuality and Victimization

As Yvonne Tasker has observed, female-centered texts frequently work to "explain away" the aggressive tendencies of female heroes in various ways. Consistent with the female heroes examined by Tasker, the imagery in *Nikita* and its extratextual material foregrounds the sexuality of the female body rather than its physical power. Julie D'Acci has shown how extratextual discourses construct and imply particular audiences.[1] Extratextual discourse thus provides a framework for how we are expected to read the text. Promotional material for the series, including ads and web pages and press coverage routinely sexualize the body of the star, Peta Wilson, and her alter ego, Nikita. Article after article details Wilson's/Nikita's provocative appearance and Nikita's sexual relationship with Michael. Most often, her physical aggression is tied to the exaggerated sexuality of the dominatrix.

For example, Wilson graced the cover of the premier issue of *Gear: The New Magazine for Men*. The cover features a heavily made-up Wilson, in

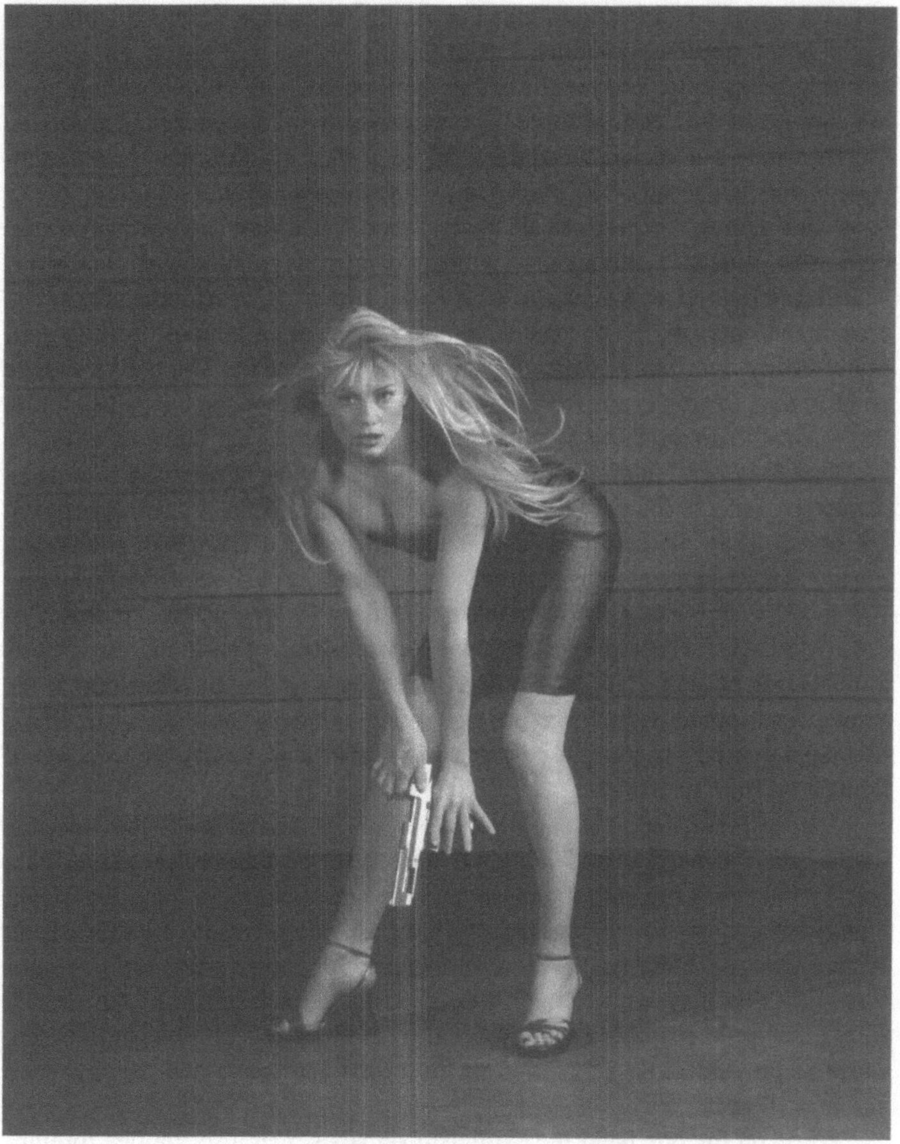

Figure 1. Peta Wilson as Nikita.

tank top and ragged underwear, with her blonde hair tousled. Despite her vulnerable appearance, the article header reads: "Dangerous Woman. In *La Femme Nikita* Peta Wilson assassinates people. When our writer interviewed her, she put him in a headlock. And the results of the photo shoot? They'll kill you, man."[2] Correspondingly, the article features a variety of provocatively posed photos, including one in which Wilson sits naked on a toilet, smoking a cigarette, and another, shot from behind and below, gazing up at her bare leg, as she climbs a tree. The wildness the photo evokes is reiterated by the article's text, which equates her to a Viking, describes her as "untamed" and "cleansed of all inhibitions," and quotes her as disdaining "pretty boys" and asking, "Where are all the real men?"[3] Despite *Gear's* obvious delight with Wilson's animalistic, violent, dangerous side, she is finally revealed as a *hunted* animal, confessing to the author: "I'm scared shitless. I'm a frightened person."[4] The article frames Wilson as an animal, lacking control over her own body, ultimately in need of protection. The active female body is thus exciting, not for its promise of power, but for the presumption that one could do anything to her in bed.

This focus on Nikita's looks and sex appeal is hardly limited to outside reviewers. Indeed, the network emphasizes not only Wilson's sex appeal but the appeal of all of the series' characters. The official USA web site for *La Femme Nikita* features a page titled "Drop Dead Looks" in which browsers can access a series of glamour photos of Wilson as well as co-star Roy DuPois (Michael).[5] These photos are pure cheesecake. My particular favorite is a full-length photo of Nikita in a tight black dress as she aims her gun at the screen, legs spread wide. From the right-hand corner of the screen, an almost life-sized gun points straight between her spread legs, as rhythmic bursts of light shoot from its barrel.

The titillation factor is highlighted within the series itself. For one, the theme music evokes sexual imagery. It consists of a heavy pounding bass line, symbolizing rugged toughness, and a woman singing a long, wavering, high-pitched note that is orgasmic in tone. She then speaks breathlessly in French. The sexiness of the double image of masculine-feminine is reinforced by this play with French. The series' title is taken from the French film, one would assume, not only because of name recognition but also because of the American perception of the exotic and erotic quality of French and French girls.

While on the one hand the exoticism signals sex, it also implies foreignness and the presence of the other. By making Nikita exotic, the female spy is positioned as essentially "different." Likewise, Nikita is the only female spy with a recurring role (with the exception of Madeline, Section's second-in-command, who does no fieldwork). The effect is to heighten the sense of

Nikita as a spectacle, consequently lessening the viewer's sense of Nikita's subjecthood and power.

The dual nature of the "dominatrix" is echoed by the duality of Nikita's function as a spy. One of the obvious ironies within *La Femme Nikita* is the discrepancy between surface appearances and reality: Image and reality are consistently revealed as never being the same. For example, there is the paradox between Nikita's outward appearance and her ability to kill. In one first-season episode, Nikita, posing as the long-lost daughter of a villainess, smiles and laughs with the villainess while drugging her tea with sleeping pills. The "lie" of her duplicitous behavior is reinforced by the "lie" of her appearance. In this scene, she wears an innocent floral print dress. She has her hair pulled back in barrettes, a look that signals innocence and youth. She is brightly lit by sunshine coming in from the window at the breakfast nook. The light, setting, and costuming signal happiness, truthfulness, and if we can judge by ads for laundry detergents, happy relationships between mothers and children. Yet the action of the scene completely contradicts this appearance. Spies often behave in just this sort of duplicitous manner. However, because Nikita is a woman, one of the resonances of this split in appearance/behavior ties into a cultural belief that women are natural actresses and liars who regularly play roles.

However, during the first season, the threat of Nikita's trickster ability is very much contained. First, the conventions of TV action series require that the protagonist cannot be a villain; therefore, we find that Nikita's ability to lie and fight is contained by the dialogue of the show, which works overtime to code Nikita as unambiguously "good." Still, although the dialogue positions her as the "good" heroine and ultimate victor, it also works to undercut the viewer's sense of Nikita's power.

The first season's dialogue consistently offers explanations for Nikita's physical skills. She must fight to protect others. She must also fight or be killed. It is noteworthy that the foremost explanation for Nikita's ability to a *take control* (i.e., her "masculine" fighting power and her identity as a "spy" and killer) is her actual *lack of control* over her own life. Removed from the protection of a society governed by constitutional rights and the protection of family and friends, Nikita must serve the agency or die. When she kills in the line of duty, she is actually fighting to protect herself rather than acting on her ideals, a situation that causes her constant torment.

It is crucial that Nikita never actually did commit the crime for which she was sentenced to death and which was the reason Section recruited her. This is quite different from the film versions, in which the heroine was, in fact, guilty of the crimes of which she was accused. Again, this is at least partly the result of TV conventions of unambiguously good protagonists, but it also connects to the way the series works to justify her use of violence and reassure the viewer that the violent hero is no real threat. Nikita's violence is further justified by the "victimized" position she occupies within the text; her status as victim will be discussed later in this chapter.[6] In addition, Nikita's killings are always justified by self-defense or presented as defending others. The depiction of female violence as maternal and protective is common to female-centered texts and draws upon notions of female moral superiority.[7] In fact, many of the episodes revolve around Nikita's unwillingness to kill.

As Nikita repeatedly states, she longs for a life of "normalcy." Despite her training as a "masculine" killer, at heart, Nikita is actually a traditional "feminine" woman, concerned with the care and welfare of others, who longs not for a life of adventure but for safety and security as well as a home, friends, and family. She works out of necessity; if she had her druthers, she would be at home. This is noteworthy because it echoes a lot of current media rhetoric about women who are "forced" to work outside of the home but feel guilty about it.

In its first few seasons, Nikita's "victimization" is signified through her visual presence. Peta Wilson is tall and quite thin with a boyish look reminiscent of a young Katharine Hepburn. On the one hand, this appearance symbolizes Nikita's "masculinization"; she looks like a man, which perhaps explains why she fights like a man. On the other hand, this body type is the preferred "feminine" body type in the world of fashion. Within the dialogue, Nikita is supposed to be a highly attractive woman. However, her body, lacking in fat, especially combined with the use of unattractive close-ups, her frequent slouching posture, and the outlining of her eyes with circles of dark or running mascara also reads as an image of a woman who has been starved and battered. During the first season she often resembles a concentration camp victim.

This image of victimization has a particular association with the youth culture audience towards which *Nikita* is marketed. Clothing and music are used to depict Nikita as a member of a specific youth culture. Nikita is a hip, twenty-something, techno-grunge goddess; her appearance and her taste in music and decor reflect current trends as seen on MTV. This is obviously done to appeal to the viewers in the Gen X subculture. However, these images signal some unsettling views about the relationships between youth culture and politics, individuality and government control, as well as feminine power.

For example, *Nikita*'s musical score is important because it not only sets up a style which is coded to appeal to the viewers of MTV but because it represents Nikita as occupying a certain "outsider" position in the cultural order, the disempowered, jobless, poor, twenty-something grungster. This connection is also made by the dialogue, her manner of speaking, the flashbacks to her rugged life on the streets, and by her choice of grunge clothing and hip hair styles. Following an MTV-style aesthetic, the music is used to create moods and attitudes. The music is grunge and industrial-band influenced. It is foregrounded more than in most action shows. As in music videos, the music often rises up to take center stage, its "attitude" and cultural associations replacing action as the center of importance.

A typical example of this occurred at the end of the series' premiere, in which Nikita listens to music while burning a picture from her past life in a candle. The scene, which contains no dialogue or action other than the burning of the picture, is essentially a music video. The scene suggests her inner torment; at the same time it aligns her with the youth culture. Her pain, expressed by the burning of the picture, is aligned with the lyrics of the music and with her grungster character. Nikita's pain and longing are both expressed by the angry music and angry appearance of the grunge subculture. She embodies grunge as the music and image embody her.

As well as recalling MTV and youth culture, this scene places Nikita in a certain political position. Within our culture, the grungster/industrial music lover is associated with angry young men and women, like Kurt Cobain and Courtney Love, who represent themselves as and are seen to be outside of the dominant ideology. Furthermore, her presence as a woman grungster, calls to mind images of female victimization and rage, like that expressed by Love and her band *Hole* and the Riot Grrrl movement. These women are not only outside of dominant ideology but as women are oppressed by it much as Nikita, an outsider at her agency, is oppressed and controlled by the agency forces. Thus, by her association with grunge, Nikita is tied to images of oppression but also to a certain kind of youth rebellion.

It is interesting, however, that this kind of youth rebellion that Nikita represents in the first season is relatively inarticulate and powerless—expressing rage but in a self-destructive manner. Like Kurt Cobain, who expressed his rage ultimately by killing himself, Nikita's rebellious rage is more image than substance because it threatens to turn inward. Nikita rarely speaks in long sentences. Her arguments with people usually consist of howled, monosyllabic screams of "Why?" She is the epitome of non-expressive youths, so worn down with the oppressive practices of life they lose articulation and, eventually, the ability to protest in a meaningful way. Rebellion is no longer done physically but through passivity and non-action.

In the first season, Nikita embodies passivity. In addition to her lack of loquacity, frequent close-ups of her pouty face reflect the look, the hopeless boredom of dissociated youth, in such fashion icons as Kate Moss or the bored models of Calvin Klein: This is the look of heroin chic. It is reinforced by Nikita's physical self, which often appears pale, sick, and traumatized. *La Femme Nikita* thus seems to be consciously playing upon images in the dominant culture of youth rebellion. These are images of rebellion through passivity, rebellion through withdrawal, where anger is expressed through clothes and music and self-mutilation/starvation rather than more direct political action.

This imagery has several effects. For one, it works against the idea of the female hero as a powerful individual. Nikita's body is either sexualized or presented as sickly, inwardly rather than outwardly focused—the antithesis of the physically robust body of the male hero. In addition, her body is shown as constrained and repressed by the external world. For example, the vision of youth rebellion depicted in the series suggests a particular attitude towards government and individualism. The connotations of rebellion and oppression, taken in context with Nikita's oppression by an agency, actually articulate anti-government themes and a deep concern over the right to bodily privacy. Nikita is not only presented as a victimized woman, but her outsider position and rebel status make her weekly confrontation with Section a symbol of a larger cultural fear of being watched and controlled by external forces.

Surveillance is a major theme. Several episodes explicitly deal with Nikita's fear of being watched. In one episode she learns that her apartment is being monitored by Section. The leader of Section, Operations, is often shown peering at the agents through his office window. Nikita is constantly warned to "watch herself." Fear of surveillance is evidenced by the series' opening credits. It begins with an opening shot of a bedroom. The camera pans along behind the bare feet of a woman walking across a bedroom floor, as Nikita, in voice-over, explains her work for the agency. The "tracking" pan shot is similar to those used in horror movies to evoke the point of view of a stalker. The camera pulls back to reveal Nikita, at the window, looking over her shoulder as the voice concludes, "if I don't do what they want, they'll kill me." There is a quick montage of Nikita fighting an attacker. Then the music begins over the next shot of a hand at a computer keyboard. There is a cut to Nikita walking along a city street as numbers and a colored bar scroll down the screen and appear to be "scanning" her body. Interspersed between these scenes are brief three-quarter view close-ups of Michael, who is positioned so as to imply he is the typist. It is also clear that he is watching and observing the body scans. The titles then switch to a variety of head shots of the series' stars. It is notable that in many of these shots, Nikita looks unhappy. The one scene in which she is smiling, she is wearing lingerie and is bent over towards the camera revealing her bosom. The credits end with another glimpse of the computer keyboard and rolls of film spread out on an editing table. The final shot of the sequence ends with a shot of a her new ID card. Even the lyrics of the opening credits play with themes of surveillance, saying "Cherchez la femme, cherchez la femme, dans la nuit, dans la nuit." (Seek/look for the woman in the night.)

Not only is Nikita watched, she is constructed. Scanned by the computer at the beginning of the sequence, she is watched and photographed throughout the opening credits. These images are apparently used to construct Nikita's spy identity. The use of computer imagery—keyboards, scanning, and ID cards—implies that technology is the primary means of observing Nikita and that technology has the ability to remove her from her life on the streets and construct her into a spy. Importantly, the images of her unhappiness also convey her oppression by this continuous observation.

These images clearly connect to theoretical observations about surveillance culture. Foucault's writings have shown us how the everyday organization and regulation of time, space, and movement train, shape, and discipline our bodies. Bodies thus become marked and constructed by the historical period's practices, which reflect ideas about identity, self, race, class, gender, and sexuality. Regulated by cultural norms, the body becomes "docile," a process which Foucault explains through the model of the panopticon. The panopticon is a prison in which the prisoners are visible at all times, subject to the gaze of the guards. The purpose is to make the prisoners feel they are under constant observation. They thus learn to monitor and control their own behaviors, whether they are under guard or not.[8]

Similarly, Nikita is under constant observation. She must learn to self-monitor her own actions in order to ward off punishment. Disciplining her body into the form of the killer that Section wants her to be, she is constructed, formed into a new being. Even so, Nikita rebels against those who watch her. She struggles to maintain a sense of her own individuality, lead her own life, and maintain her privacy. The ongoing theme of the fear of surveillance and bodily control is made horrifyingly concrete in a late third-season episode in which Nikita discovers a secret nursery hidden within the bowels of Section headquarters. It appears as if Section may have gone so far as to *clone* her, using their scanning devices to actually construct a new, more obedient Nikita. Even her mental autonomy is invaded. In "Under the Influence," she learns that she has been brainwashed by Section into falling in love with a vicious psychopath and that her apartment is riddled with pro-Section subliminal messages. In "Outside the Box," Nikita discovers that she may or may have been set up by Section for the crime for which she was accused in order to work for them. The third-season finale, "On Borrowed Time," ends with Nikita again being brainwashed by Section.

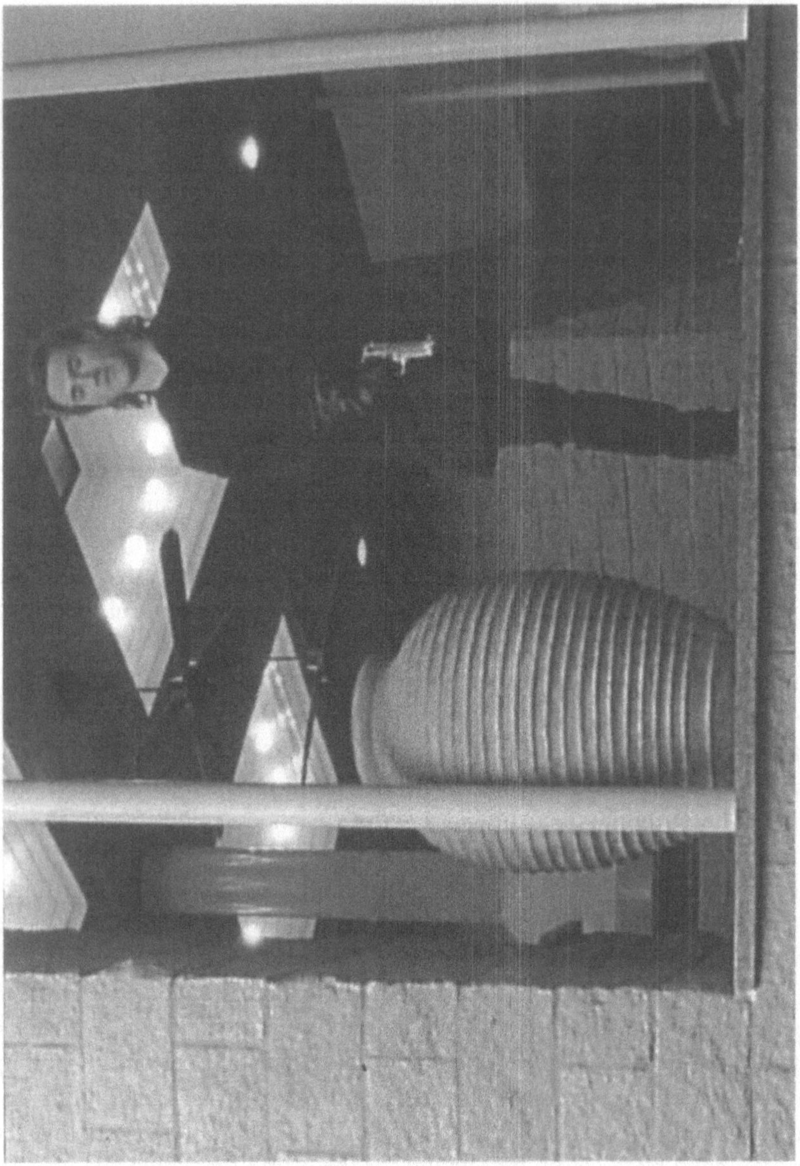

Figure 2. Michael in Section headquarters.
La Femme Nikita ©1998 Warner Bros. All Rights Reserved.

The show's concern with loss of privacy is seen in the way it contrasts the individual with the group. The world of the agency, symbolized by Michael's spare black garb and the cold mechanics of the "control" room are juxtaposed to Nikita's own "rebellious" style of dress and playful manner of decorating her home. Similarly, in one episode, this distinction is created by dramatic lighting. After tracking down a killer, Nikita sees him smoking, drinking, and laughing in Operations' office. Enraged by Operations' betrayal, Nikita threatens to shoot him as she stands in a dark room surrounded by a circle of bright light. She does not shoot, but puts her gun away and walks from the room. The circle of light remains, signifying her individuality and separation from the corrupt agency.

The use of "light" to represent individuality and "dark" to represent Section is interesting because it connotes value judgments about individuality versus organization which are upheld by the rest of the text. In the world of *Nikita*, individuals are embattled and endangered while organizations are necessarily corrupt. Although the agency is described as having "just" ends and does defend against nuclear terrorists, white slavers, and drug lords, its "means" are actually quite brutal. Section uses torture as an interrogation method and kills its own agents as well as its prisoners. The ambiguity of the "goodness" of agencies is interesting mainly because the show's view of the various agencies/villains seems to be so flexible: It suggests a certain relativity to the morality of their actions. In fact, what becomes clear is that, working for just ends or not, *organization* itself is the problem. Even a "just" agency is ruthless. It is a world in which the constitution is ignored, and individual rights are nonexistent.

Nikita presents a world in which the integrity of the individual body is under constant threat. This clearly places Nikita, like many heroes, within a tradition of American individualism. Many male-centered texts depict a similar distrust of organization. *The X-Files*, for example, similarly depicts government agencies as corrupt, capable of harboring secret sectors in which constitutional individual rights are unimportant. Such texts connect to a cultural current which questions the power of the government and of technology in our lives.

Nikita's depiction of surveillance culture reveals Nikita to be a "docile body" shaped by the gaze of Section as she learns to be the spy they want her to be. Over the years, Nikita's appearance evolves. Once a street punk, she begins to conform to the stylish, sophisticated black garb of the other Section operatives. Madeline trains her in the use of make-up. In her various missions she is shown adopting a wide range of costumes. In addition to her evolving appearance, she becomes a more self-actuating heroine partly because she has internalized the values and skills of her environment. In the first season, Nikita always told the truth—at least to the audience, who are privy to her point of view and private reactions. In the second season, Nikita begins to lie and/or hide and manipulate the Section's system.

For example, in "New Regime," Operations, the leader of Section, is hospitalized after an assassination attempt. Nikita is picked to be second in command to the new section head. The climax of the episode occurs after a meeting between Nikita and the new head of Section. She convinces him that she will kill Operations for him. This is the first episode in which the viewer is *not* made aware of Nikita's point of view and every action. At the end, the viewer learns that Nikita has been working for Operations all along.

Her growing ambiguity as a character allows her greater control and power within Section. By the second-season finale, "Endgame," she even becomes a chief player in the continuance of Section. With the aid of the former head of Section, Adrian, Nikita begins to spy on Section. She gains access to restricted files and corridors. Meanwhile, she begins to lie to those around her. A major subplot is that she is now outside of Michael's help—and his control. For the first time, Nikita holds power over the leaders of Section as well as over her own destiny. At this moment, Madeline and Operations realize that Nikita is a threat but that her actions and popularity within Section have made her too powerful to kill outright without more agent rebellions. By the middle of the third season, Nikita has grown into a force with which to be reckoned. Even though she is still being spied upon, she has to some extent achieved the power to reverse the gaze, to look back at Michael and her superiors. Frequently, as in "Endgame," "Outside the Box," "Playing with Fire," and "Walk on By," she and Michael are engaged in deceptions against Madeline and Operations, in which Nikita's ability to dissemble is called upon.

One notable change in Nikita's behavior is that her unwillingness to kill gradually dissolves. While the first season saw her never shoot until she was shot at, by the second season she kills in the course of duty without seeming too worried about each individual death. Not only does she kill on the order of Section but for her own purposes. In "Old Habits," for example, Nikita is enraged by Section's protection of an informant who also happens to be a sadistic serial murderer. The informant refuses to cooperate with Section unless they give him a prostitute whom Nikita has been protecting. When the prostitute disappears, Nikita believes Michael turned her over to be killed. He denies the accusation, but it is unclear whether he is being truthful. Although justified by the informant's bloodthirsty behavior, it is notable that Nikita, in contrast to her first-season behavior, is quite willing to kill him.

When the informant is killed by a member of the enemy terrorist group. Michael believes that Nikita arranged the hit by giving him up to the other side. In a scene that parallels their earlier confrontation, Nikita innocently denies involvement. The quick reverse shot to Michael's concerned face tells the viewer that Michael believes she is lying. This reaction shot is imperative because it creates a sense that Michael, the master liar, is being hoist by his own petard, gazing at a monster he has helped create. Although Nikita's motivations are perhaps understandable, it is more important that her devious, murderous actions and blank facial expressions replicate the way Michael, Madeline, and Operations behave on a consistent basis—she has internalized the rules and practices of Section's culture.

In the fourth-season finale the audience learns alongside a stunned Michael, Madeline, and Operations that Nikita has actually been working undercover for Oversight, the organization that oversees Section, for three years. At the conclusion of the series, Nikita learns she is the daughter of Mr. Jones, the mysterious head of Section. Taking his place as head of Section, she sets Michael free to be with his son and presumably makes the system more humane. The ending is bittersweet; Nikita clearly desires to be with Michael but is forced to assume her position for the greater good. Although triumphant and powerful, she remains stuck inside the system that separates her from the life she wishes to lead.

Importantly, Nikita's power within the organization and environment that once oppressed her is linked to her ability to adapt her appearance, adopt different roles, and hide her actions and motivations not only from other characters but from the viewer. The fact that so many of her later actions take place off-screen and her motivations are hidden from the audience means that Nikita cannot be read merely as an object of sight despite her frequent sexualization. She has moved outside of the "three gazes" outlined by Mulvey. Invisible to the camera, other characters, and the audience, Nikita steps outside the power of the "male" gaze. Her body is not presented as a static object, pinned by the gaze, but is a more fluid figure.

The sense of bodily fluidity created by camera work is echoed in the series' narrative. The depiction of Nikita's deception and role playing as the route to her ultimate victory underscore the power of physical transformation. By the end of the fifth season, she has become plastic, transforming herself at will to fit the situation. As Susan Bordo has noted, the idea of the body as plastic and mutable, as a "playground," something that can be shaped and molded through willpower, is a commonplace attitude in this postmodern era.[9] It is also a central tenet of male-centered texts, which regularly highlight the hero's control over his physical self.

However, plasticity implies one is not limited by material constraints at all—one can do or be anything that is desired, it is the mind and will that are the most important. In contrast, Nikita manipulates her body not by choice or out of a sense of play, as do male trickster heroes like James Bond or *The A-team*'s Face, but in response to her dangerous environment. Moral ambiguity and deception are the rules of Section, and Nikita internalizes these lessons to survive. Thus, although Nikita gains great power, defeating Operations and Madeline at their own game, she is first and foremost a body that is acted upon: She changes her bodily practices such as behavior and appearance in response to the actions of those around her. Rather than reaching out to alter her world, Nikita is penetrated by the external, material world and shaped by it, a docile body. Although she will presumably work to make Section a better place, the final episodes show her still rooted within the system, needing to follow the rules which require her to live separately from Michael and disconnected from the everyday world in which she desires to live. *Nikita* transforms herself to meet the demands of social acceptability. She is more like Cinderella dressing for the ball than a trickster, changing on whim and for her own purposes.

Obviously, the text's suggestion that social power is achieved through the modification of bodily appearance and behavior has a resonance for women, whose daily lives are often shaped and governed by a social reality in which their physical selves are routinely surveyed and judged. In this way *Nikita* highlights the importance of maintaining an appropriate surface appearance and consequently underscores the importance of the body. While the male hero changes the outside world but does not himself change, his hard body symbolizing the stability of his identity, Nikita undergoes a process of transformation in which the signs of the body—her behavior and appearance—change in response to the demands of the external world. The plasticity of her body is both a sign of her lack of dominance over others and her environment, but it is also the means by which she achieves power. If the imagery of male-centered texts suggests the hero is impenetrable and hard, piercing through the world, the female hero of *Nikita* is like Silly Putty, soft, permeable and picking up the images pressed upon her.

Intimate Investigation: Visual Style and Romance

At the same time the superficial, performative qualities of the body are highlighted by Nikita's transition from victim to leader, other textual features of *Nikita* also emphasize the body and those qualities we tend to associate with the body such as sensation and emotion. In fact, although *Nikita* is an action series, there is very little fighting. Instead, the majority of the "action" in *Nikita* is its atmosphere, character development, and plot. *Nikita*'s textual emphasis on love, relationships, and being highlight the sensuality of the body. The unique visual style and soundtrack likewise fetishize the human body, both male and female, as well as bodily sensation and experience.

Nikita clearly relies on many conventions drawn from "female" genres like romance, in which the action/adventure elements have been blended if not supplanted. One could argue that the presentation of a transgressive heroine has been subverted by this tactic. Conversely, genres like romances and soap operas have traditionally allowed women to be at the active center of the narrative; their use creates a space whereby a female can easily enter the action genre. Either way, *Nikita* offers an interesting approach to action adventure that seems to signal a shift in television conventions towards a more "feminine," if not "feminist," aesthetic.

In *Television Culture*, John Fiske has argued for the existence of gendered texts. While viewing texts as "masculine" or "feminine" is problematic for various reasons, Fiske's formula nonetheless provides a useful framework for delineating differences between certain kinds of texts. "Masculine" texts feature closed narratives with clear-cut villains and heroes. "Feminine" texts, on the other hand, are narratively open. The soap opera is an example of an open "feminine" text, both narratively as well as visually. Storylines continue and evolve over time but have no final resolution. Even when characters die, there is always the possibility they may return. Sandy Flitterman-Lewis has shown that because the characters in soaps are so open to change, soaps actually encourage multiple identifications with characters.[10] As Mary Ann Doane argues, this kind of multiple identification is considered a "feminine" reading tactic.

While many TV texts today have an ongoing storyline, soap operas are also somewhat unique for their "open" visual style. Traditional filmic techniques of the narrative cinema, such as the shot-reverse shot rule, are continually broken, dislocating the viewer from a sense of appropriate space. Flitterman-Lewis argues that the soap's multiple narratives are echoed by the variety and complexity of the kinds of shot set-ups, such as variation of type of shots, camera angle, character and camera movement, which work to disperse, fragment, and amplify the spectator's vision.[11] For example, camera distance, tracking shots and zooms, create a perpetual fracturing of space. There are rarely repetitions of a single set-up; as a result, space is frequently redefined, revealing a different angle, or a different portion of a character.[12]

Flitterman-Lewis sees a parallel between what we are shown in soaps and how we see it. She argues that the constant movement of the eye that is created by these filmic techniques implies that, unlike the "masculine" gaze of classic cinema described by Laura Mulvey, there is no "'unifying presence' at the site of spectatorship."[13] In addition, soap operas are feminine in that they are highly concerned with emotion versus narrative. This can be seen in their extreme emphasis on close-ups. Close-ups are paramount, since close-ups allow an intense investigation and surveillance of the emotion being suffered. In her study of *All My Children,* Flitterman-Lewis talks about the use of the close-up:

> To amplify the subtle conflict playing across [the actress's] face....The entire sequence...could be described as "the drama" of Natalie's face as minimal dialogue (for her) and maximum facial expression intensify the narrative significance of the moment.[14]

In other words, the drama at the moment is the emotions of this particular individual at a particular, individual moment. A sense of the importance of imminence is created.

In addition to making the face the center of the drama, frequent close-ups give the viewer time to look for the "truth" of the character, to intimately survey and investigate her or his face and emotions. This search is paralleled by soaps' tendency to revolve around a mystery of some kind or another, such as the mystery of paternity, in which it is necessary for both characters and viewers to discover "the truth" of a character's feelings or actions. While this may seem as if it replicates the kind of surveillance found in "male" genres like those examined by Mulvey, I would argue that there is a qualitative difference in the kinds of surveillance that occurs in films versus soaps. The difference is one of intimacy.

First of all, it is important to note that when discussing television texts, it is questionable how closely one can follow Mulvey's formula, since the conventions of filming are different. As E. Ann Kaplan has argued in her examination of music videos, which like soaps feature a disjointed editing style, the image that is most often featured—that of the video star—becomes the unifying presence of a disjointed filmic world. Similarly, close-ups create a clear sense of a relationship between the viewer and the characters on the screen. People identify with soap opera characters and other images filmed in close-up because they "ground" the action and provide a frame of reference with which to interpret the text; it is the face and emotions which center the visual world. Because of this, the surveillance of the face is intimate in effect rather than replicating the sense of distance implied by "the male gaze."

Just as soaps create a sense of intimacy through their visual style, romances also create a quality of intimate investigation. As Tania Modleski has pointed out in *Loving with a Vengeance*, contemporary gothic romances are clearly linked to women's experience of danger and uncertainty within their domestic lives. Gothic romances feature a mystery, often occurring within domestic space, which revolves around an ambiguous character whose true nature is not revealed until the end of the story: the romance hero. The hero is an enigma, often appearing villainous, until the heroine solves his mystery by learning to accurately interpret his often harsh-seeming actions as expressions of his care and concern.

Modleski argues that this scenario appeals to women because women's real economic and physical safety has often depended on being able to interpret men's desires, thoughts, and feelings—often before the men themselves know what they want. And, as Janice Radway has observed, the heroine's discovery of the "truth" of the hero's love often coincides with her recognition of the truth of her own feelings and marks her transformation from adolescent to her "new" identity as a mature woman. Only by solving the mystery of the Other, can the heroine define herself and essentially find herself. Once again, this discovery is one that is marked by the intimacy of self-discovery, rather than distance of the discovery of the other.

It is clear from observing the notes from viewers posted on web pages devoted to *Nikita*, such as the official USA site, that viewers are watching the show in large part because of its romantic themes.[15] Certainly the ads for the series focus almost exclusively on the sexual relationship between Michael and Nikita. As one third-season promo states, "Lips that search the soul, eyes that seek the heat, this one is on fire. Beyond intensity...*La Femme Nikita*."

Nikita follows several traditions associated with the "feminine" genres like soaps and romances. It utilizes romantic plot devices and a soap opera visual style that create feelings of intimacy at the same time they stress emotions and the importance of the moment. The major "mystery" for Nikita, as in soaps and romance, is the internal mystery of relationships and the truth of human nature. In addition, *Nikita's* slightly disjointed music video editing style recalls the visual style of soaps and music videos, placing a primacy upon close-ups which in turn makes a spectacle of emotion.

The centrality of emotion seen in the series' visual style is echoed in *Nikita's* foregrounding of the romantic relationship between Nikita and Michael in the narrative. Not only is their relationship a focal point for many of the individual episodes and a major theme in the overarching narrative thrust, but it is emphasized by the extratextual elements like ads for the series or the series' web pages, which highlight the sexual tension between the characters as well as the sex appeal of the actors.

Michael is structured as the enigmatic romance hero whom the heroine must decipher. The question of whether or not Michael loves Nikita, his motives for behaving as he does, is a major theme. At times he behaves romantically towards her; at others he snubs her. Nikita tells Michael she has no idea who he is or what makes him tick. She repeatedly asks him "why" he acts as he does. At the same time that Michael is coded as "ambiguous" and is clearly seen as such by Nikita, his motives are fairly clear to the viewer. Romance novels often feature "clueless" heroines, who interpret the action of the hero as anger, while the reader, who is more knowledgeable about the conventions of romance, correctly interprets the hero's actions as a sign of his love. Similarly, while Nikita herself may have doubts as to how Michael feels about her, the audience has few doubts, particularly as the series moved into its second season. The text begins to fluctuate between various characters' points of view, an approach that, as in soaps, allows multiple viewer identifications.

The first-season finale and second-season premiere are prime examples of Michael's function as a romantic hero. In the first-season finale, Operations and Madeline decide to cancel Nikita. They assign her to a suicide mission. Michael appears to agree with them but secretly gives Nikita a communicator and tells her to use it if problems arise during the mission. While the rest of the crew is told to maintain their positions and are blown up, Michael signals Nikita, telling her it is a trap and that she is free to escape.

The second-season premiere opens with a montage of Michael as he mourns the loss of Nikita. Filmed in music video style, the music, a sad love song, plays while Michael stares off into the distance. The montage is about his "remembering" Nikita in all her beauty. During this sequence, Michael is involved in a training session of martial arts. He is forgetful and has to be reminded to start fighting. He casually breaks a trainee's leg because he is absent-minded or angry. Meanwhile, Madeline and Operations comment on the fact that his behavior and performance have gone downhill since Nikita left, and they wonder if he will ever recover.

This sequence is highly sensual, featuring extreme close-ups and with the musical soundtrack foregrounded over the atmospheric noise. The speed of the frames is decreased to create a slow-motion, blurred effect. Although Michael's face, as usual, is expressionless, the "remembrance" montage allows the viewer to enter his mind: We know that he is obsessing about Nikita. Moreover, her loss has left him enraged and emotionally bereft. He has lost his ability to function without her. In this way, the importance of the female—her emotional effect on the man—is made clear. Like romance heroines, Nikita's "power" depends strongly on her ability to affect him.

The remainder of the episode follows a typical romance pattern of the hero violently breaking down and obliquely confessing love through a sexual encounter with the heroine which is physically violent in form. Michael eventually tracks Nikita to where she is hiding on a boat. It is unclear whether or not she recognizes him in the dark. In any case, he tricks her and overpowers her, wrestling her to the ground. There is a close-up of his face as he stares into her face. Then he passionately kisses her. He breathlessly tells her "I thought I lost you." She replies angrily "You never *had* me," before violently kissing him back.

As is the case in many romances, this explosive love scene shows the emotional breakdown and rupturing of the hero but ultimately reveals the physical domination of the heroine by the hero and of her transformation to submissiveness. For example, in this episode, Nikita and Michael's love scene is lit so that Michael is in the darkness, and Nikita's naked back and buttocks are brightly lit. He is in a physically dominant position. Not only does he wrestle her to the ground, but as he kisses her, he bends her backwards over his arm. Like many romance heroines, Nikita afterwards becomes submissive, gazing up at Michael like an adoring puppy. Michael stares morosely out of the window, as Nikita asks him how he feels and if he's OK. Considering *she* is the one who has been believed dead for the last six months, this would be a more appropriate question for Michael to ask.

The show presents relationships, particularly romantic relationships, as enigmatic and fraught with danger. This dance of closeness and distance is not limited to Nikita and Michael. By the second season, everyone in Section appears to be obsessed by her or his own sex life. The serialized nature of the show and its concern with relationships clearly demonstrate that *Nikita*, while theoretically an action text, is far more easily read as a soap opera. In addition, the show's other main themes of mystery and danger may further be read as connecting to romances and soaps. The recurring theme of surveillance illuminates the show's relationship to these two genres.

Laura Mulvey has pointed out that in classic films like *Vertigo*, the metaphor of the spy is used to give an excuse for the voyeuristic behavior of the protagonist. Nikita's "spying" is a metaphor for her learning to read the "truth" of others' motivations and feelings. The fact that she is clearly positioned as the one doing the looking implies a level of voyeurism. Indeed, John Cawelti and Bruce Rosenberg argue that the appeal of the spy genre, from which *Nikita* is drawn, may be due to the fact that spies offer a fantasy of power.[16] Spies have the power of invisibility, to see without being seen and to act without repercussion. Nikita's subjection to the control of Section is likewise reminiscent of other "spy out in the cold" texts that feature lonely and isolated heroes. However, the series' narrative and visual style create a quite different kind of spy than the heroes of John Le Carre. The basic principle of the voyeur is that the voyeur is not himself seen. In contrast, at the same time Nikita surveys and regards others, she is herself constantly watched.

As earlier discussed, the surveillance imagery in *Nikita* works to contrast the oppressed individual and the surveying state. However, it can as easily be read as expressing a real-world dilemma for many women, which is also expressed by soaps and romances. Just as in the panoptic world there is no escape from surveillance, the terror of the gothic romance is effective precisely because the threat comes from within the home and the people who are most closely connected to the heroine. There is no safe space to hide from the threat. Similarly, Nikita's "home" is the paranoid and dangerous world of Section. Her "family" of co-workers is extremely ambiguous. This is a world in which very few people can be trusted, and certainly no one with power can be trusted.

Like the heroines of gothic romances, Nikita must navigate the ambiguities in her intimate daily existence. Her survival relies on her ability to understand the motives and emotional "truths" behind people's facades. *Nikita* continually foregrounds the fact that understanding is a matter of life or death. Nikita faces death if she makes mistakes in her missions and fails to understand the true motives of the villains. She also faces death if she fails to act appropriately towards Operations or Madeline. She is constantly placed in situations in which her survival is dependent upon her ability to make the right choices. Like the romance heroine, she finally needs to rely on instinct. Her choices are almost all emotional, based on reading the personality or desires of another.

For example, the episode "Recruit" revolves around Nikita's assignment to train a female agent and determine if she is spy material. The stakes are high: If Nikita decides not to pass her, the new agent will die. The character of the trainee is a mystery. When Nikita announces her decision, she is stunned to learn that the trainee has already been killed; the real test was not of the trainee but of Nikita's ability to read personality. Had she failed in this, she would have been killed.

The way Nikita is surveyed at the same time she surveys, a move that breaks down the sense of distance between subject and object, particularly when combined with a narrative focused on relationships, creates a very intimate atmosphere. While less fractured than that of a soap, *Nikita's* visual world creates a similar sense of intimacy, which, while concerned with the discovery of "the truth," stresses emotion. Visually, *Nikita* has a very distinct style created by the frequent use of techniques such as close-ups, slow motion, and blurred photography and also features a prominent soundtrack that is often foregrounded over dialogue and atmospheric sounds. These techniques underscore the importance of emotion as not only *part* of the narrative but actually the *drama* itself. Frequent close-ups draw attention to the actors' faces, and the use of slow motion highlights the sense of the *moment*, especially the emotional content of the moment, as being incredibly important. Slow motion close-ups, like sports replays, tell the viewer to pay attention because this moment is important. In a sense, the text invites the viewer to not only pay attention to the moment but to study it, re-hash it, reconsider it.

The invitation to examine and reexamine the individual moment may also be said to be "feminine" in that it recalls the constant evaluations and revaluation that women do in regard to the enigmatic male. It suggests that, although there may be no one meaning, if one tries hard enough, focuses on enough small details, it might be possible to discover the "truth" about a person's motivations. Thus, visually, the use of close-ups and slow motion echo the themes of enigma and investigation that are at the narrative heart of the show.

Television has long been considered an intimate medium since it takes place within the home and features a "talking head." Despite this, many television genres, such as action/adventure and sit-coms, rarely utilize the close-up. Most sitcoms replicate the theatrical stage, with actors filmed in full body view. Action series also feature more long shots in order to see the physical action. Dramas obviously rely more on close-ups. However, they tend to show more waist-to-head shots. With the exception of soaps, the use of extreme close-ups is actually quite rare. *Nikita*, however, frequently features extreme close-ups. This becomes a particularly interesting aspect to examine when we consider that the series is an action drama. Yvonne Tasker has argued that it is the lingering visual image that most affects viewers' perceptions. The audience's lasting impression of the hero stems from the visceral effect of sight of the body in motion. In *Nikita*, the lingering image is that of the emotive face.

Frozen Emotion

Certainly the visceral quality of *Nikita*'s visual style is its most compelling and memorable aspect. However, it is still interesting to note that from a visual standpoint, there is actually very little *action* in this action series. Action sequences are always a means to an end, rather than the end itself, in that they are used to enhance or stimulate the real action of the series: the characters' relationships. Therefore, the visceral moment the viewer is likely to recall is not a moment of physical action but a close-up moment of emotion.

The action sequences that occur frequently fetishize both male and female bodies but suggest the superiority of the male action body. In addition, they function primarily as ways to communicate a concern with emotion, sensuality, individuality, immediacy, and time. As previously noted, *Nikita*'s more obvious textual characteristics are the use of slow motion, music video editing, music, and blurred frames which set off the action sequences. In this way, moments of action are visually treated as important, set aside, and as

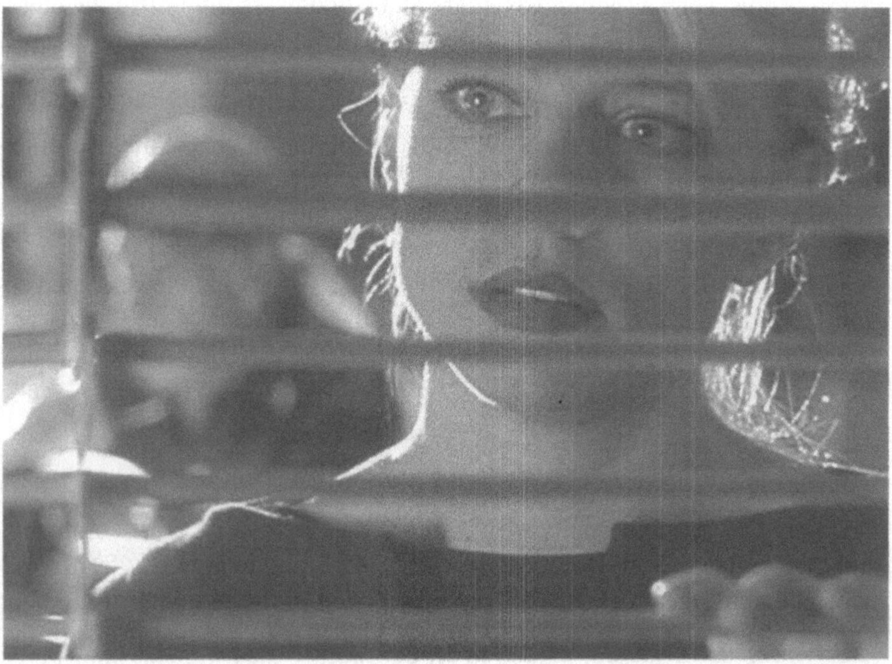

Figure 3. Nikita is both surveyor and surveyed.

outside the "normal flow of time." While this kind of framing occurs elsewhere, such as in highly emotional moments, it is featured in almost all action sequences. In addition, the sequences are sexualized, both by motion, costuming, and the use of music.

For example, one episode begins with a sequence in which Nikita and Michael break into the office of a bad guy, conveniently housed in the basement of a disco. This setting gives Nikita the excuse to dance across a nightclub floor to Michael. He puts her hand on his belt from which she draws a gun; then she walks downstairs and into a computer room. It is not clear why she didn't just carry in the gun herself except that way there would be no satisfying sexual innuendo. Then Nikita simply inserts a disc in the computer. When she is discovered, Michael comes to her rescue. The action sequence now switches into high gear. Michael is filmed in slow motion as he kills several people, jumps down the stairs dramatically, and is eventually captured. Nikita does little more than duck behind a table and fire her gun.

Nikita's role in the action sequence is sexualized, as she is limited to dancing, groping Michael, and walking with a gun. Yet it is Michael's body which is fetishized. Not only is the camera angle low, emphasizing his groin, but slow motion allows the eye to linger over his every motion. This works not only to provide a moment clearly marked as outside the normal flow of time in which one can admire him as a sex symbol but also gives a sense that this man's character and soul can be distilled in these few short moments. This kind of slow-motion glorification of Michael occurs quite often but is rarely done with Nikita. As in this episode, Nikita's body in motion typically appears as a dancer or in flight, fleeing from danger, while Michael's body, though fluid and sexualized, is nonetheless fighting aggressively rather than responsively.

Conversely, Nikita's character seems to be defined less by what she does than by *how* and *whom* she does it with: Concerns with narrative and relationships fill the bulk of her on-screen time and take precedence over her moments of fighting. So although there certainly are moments when Nikita does fight, the fighting is downplayed and not typically glorified as it is with Michael. When her fighting ability is stressed, as in "Looking for Michael" when she is double-crossed by Section, the action is not slowed down, and she is fleeing. The male agent remains the deadlier of the species.

Nonetheless, I think that the manner in which the action sequences are filmed should not be dismissed because they highlight the expertise of the male supporting character more often than the star. Bearing in mind that the action sequences replicate other moments within the text when music, close-ups, and slow motion are used, the general thrust of the text seems to be aimed at saying "Stop. Look. Listen. This moment, slowed down, frozen, emotionally close and visceral, is important," suggesting that one frame can be more important than the whole of the narrative.

Moreover, the frozen moment is one concerned with either emotion or sensuality. Likewise, the foregrounded sound track enhances the sensuality and emotional content of the moment. The sound track features the sounds of techno and grunge, with such bands as Depeche Mode, Garbage, Vibrolux, Morphine, Enigma, and Hednoize. While, on the one hand, much of this music evokes the nihilistic worldview of ravers, it is also a highly sensual music that is based on beat and rhythm. Maria Pini, in her study of female rave goers, found that the beat of such dance music ties listeners into a sense of imminence, of being "in the moment" among a community in which the boundaries between self and others are dissolved.[17]

At the same time the beat is heavy and overwhelming, as in Garbage's "Number One Crush," the voice is very often partly submerged by the musical instruments. This actually creates a sense of intimacy, as if you are in a loud club, with the singer whispering in your ear. The dance music features a beat that resists closure and, like the use of visual fade-out, suggests a continuance of the music after the song has ended.[18] It thus has a never-ending, eternal quality to the sound. Susan McClary describes such "feminine endings" in music as "creating...narratives that celebrate multiple rather than unitary identities, that are concerned with ecstatic continuation rather than with purging and containing."[19] The quality of sensuality and continued time evoked by the series' visual style is echoed by the musical score. This, when combined with the heightening of the moment, further creates the sense of an eternal moment, frozen in time as it also extends into time.

The way *Nikita* focuses on the importance of the moment creates a wonderful sense of imminence, of presence, of being in the moment, that is extremely interesting. At the same time the viewer is drawn into the moment, the visual style calls your attention to it, rupturing the moment. It thus creates a feeling of the intangibility of the tangible moment and of something existing beyond what is spoken or known. Further, there is a great tension between the structure of the narrative—the desires, secrets, and gaps between people that keep them separate from each other—and the desire for unity and sensuality expressed by the music.

In addition, the slowed-down camera speed, which draws attention to movement, emotion, and life by slowing time down, actually parallels the show's narrative concern with life or death. A theme that is repeated throughout the series is the idea of staying alive. The agents fight against being killed by their superiors; with every mission, they risk death. There is also the theme of "life" as expressed by Nikita, as being more than just life; it is about "living" or having a "life" of her own in which she is happy and free. At the same time, movement, through the use of MTV sound and filming style, becomes a kind of poetry, evoking emotions on its own terms.

Valorization of the Feminine

The focus on emotions, relationships, sensuality, and the use of "open" narratives in *Nikita* mark it as a "feminine" text that also valorizes many "feminine" if not "feminist" ideals. For instance, while romances reinforce ideas such as power being dependent upon the heroine's connection to a male, they also stress the importance of the "feminine" in life. In fact, one of the main themes of romance is how women are able to distance themselves from their "natural" victimized status as women, gaining power and subjectivity in the process. Because of this, it has been widely argued that romances are the equivalent of male adventure/journey stories.

Thus, within the romance tradition, the heroine is actually quite proactive in the sense that she manages to substantially alter her environment. For example, she often enters into a mysterious household or one which is antithetical to "feminine" values: The law of masculinity holds sway. Over the course of the novel, the environment is made softer, friendlier, and happier through the efforts of the heroine. Similarly, by the third season, Nikita has gained a small coterie of friends who begin to work together against their oppressive bosses. Nikita's "feminine" association with morality, caring, and communication is, of course, rather essentialist, and it is disappointing that Nikita gathers a group of (male) friends, largely, it appears, because she is pretty. Nonetheless, due to the fact that she is in many ways so powerless, her resistance in the face of overwhelming odds may be read as displaying her extreme moral strength.

Likewise, the secretive culture of Section is presented as pathological. Michael, who has obediently learned to follow the rules, is miserable—a tortured, tragic hero. The world of secrets and silence likewise endangers and oppresses Nikita. However, in contrast to the silent, self-controlled Michael, she strives for caring and open communication. Despite her "feminine" ethics, her anti-Section stance actually is about individualism. Contrasted to the "normalizing" Section gaze which requires her to leave her "true" (caring) self behind, she is seen as battling for her individuality and personal control over her own life.

Indeed, this seems to be a chief interpretation of her character, judging by some of the fan fiction and Internet chat on sites devoted to the series, which also reflect an appreciation for the series' romantic strain. For example, "The Cat" is a work of fan fiction written from Michael's point of view. He ponders that Section had taught them both that

> If someone had a weakness, you exploited the weakness....The difference was, she still believed she could somehow bend those inflexible, iron forged maxims if it meant hurting someone she cared about. Always find a way *around* it. Somehow. Michael wished he had that freedom, that luxury. That *strength*.[20]

He later acknowledges that if Nikita changed to his way of thinking "the only part of him that still existed would cease to breathe, and he would be a dead thing too."[21]

This is a particularly potent example of how viewers interpret the series as romance. Importantly, the appreciation of her strength is based on a valuing of a "feminine" ethic. Nikita is strong because she resists the (male) culture of dominance existing within Section in favor of protecting (female) relationships. Notably, "The Cat" portrays this contrast as also being between the values of life versus the culture of death.

The difference in attitude towards dominance, life, and death, is one of the chief distinctions between heroes and heroines. At the same time Nikita is an individual fighting external forces, she is also framed as valuing relationships and connections to others. Unlike the classic hero, who tends to be concerned with controlling others, the heroine seeks to protect her body in a dangerous environment that seeks to remove her self-determination.

Conclusion

Nikita utilizes filming and narrative techniques and codes drawn from "feminine" forms of mass culture such as romance, soaps, and dance music. Like romance novels and soap operas, *Nikita* suggests that communications and relationships are of the highest import. Relationships—ongoing and within individual stories—are always central to the narrative and are given more visual attention than the supposed action sequences or the plots of individual episodes.

The romance narrative in *Nikita* is not fully resolved. In sustaining the enigma of the other, the show plays with romance narrative but actually shies away from the happily-ever-after resolution that recuperates the heroine into the patriarchal order.[22] Instead, there is an emphasis on wondering, questioning, and emotionally exploring, which seems to signal a more "feminine" if not feminist type of text. Like soaps, *Nikita* is addicted to analyzing emotion, addicted to the intensely pleasurable moment of intimate, personal revelations, but the revelations are always temporary and fluid, the tension between characters remains, and numerous high points of emotion are able to stay in play over and across episodes and seasons. *Nikita* features a definite progression *forward* for each of the characters. As in soap opera viewing, a sense of history emerges, and the interpretation of the text becomes highly dependent upon the viewers' previous knowledge of the characters and their motivations.

Certainly, Nikita is often objectified and functions as sexual eye candy. However, more interesting is the fact that *Nikita* portrays its hero as an individual who is oppressed, surveyed, and constructed by organization and society. Nikita's only way to survive in the world of surveillance is to learn to begin to play the game herself, to become more flexible and plastic, and to transform herself into various roles. In a sense, she becomes a postmodern heroine, whose identity remains unfixed and unstable, whose identity, if she has one, exists only in the moment, a moment between life and death, thought and emotion, a moment which can never be fixed or contained.

CHAPTER FOUR

Aeon Flux and the Grotesque Body

Aeon Flux may be thought of as "the mother" of the trend of television female action heroes that began in the nineties. Created by Korean-born, American-raised artist Peter Chung, *Aeon Flux* first aired as a series of short films on *MTV's Liquid Television* in 1991. The show was adapted into a half-hour series in 1995, prior to the appearance of *Xena: Warrior Princess*, and two years ahead of *Buffy the Vampire Slayer. Aeon Flux*'s presence on MTV guaranteed Aeon a unique position in the emerging pantheon of female heroes. MTV has had a pronounced effect on contemporary visual culture, influencing both media style and content. The network's trend-setting reputation and appeal to youthful audiences garner its shows both critical and popular attention.

With its introduction on *Liquid Television*, *Aeon Flux* quickly became a cult classic. The character of Aeon Flux was named to Esquire's 1996 Top 20 list of "Women We Love Even Though They Can Kick Our Ass" and *Time Magazine's* 1996 List of Animation Winners.[1] In addition, the series has spawned a video game and a 1995 book.[2] Written by two of the series' writers, *The Herodotus File* details Aeon's recruitment as a double agent.[3] All of the original short films as well as the series' full-length episodes are now available on a series of video cassettes.[4] A live-action film version is reputedly in the works.[5] *Aeon Flux* is of particular interest because its weird, animated imagery is at odds with the appearance of typical television series and has won the series critical acclaim.[6]

Since its inception, MTV has been considered a bastion of male dominance, a breeding ground for hegemonic and sexist imagery.[7] At first glance, *Aeon Flux* fits into the visual system of the network. Scantily clad in leather thongs, thigh-high boots, and a chastity belt, Aeon is a dominatrix designed to titillate MTV's youthful, male audience. The sexual nature of her presentation suggests that, like other sexualized images of "powerful" women, the series is paying lip service to feminism while conveying deeply masculinist attitudes in drag. However, in spite of her obvious sexual appeal, Aeon is never pure spectacle. Something about her character and function in the narrative remains firmly outside the "gaze" or indeed, any narrative resolution.

Series creator Peter Chung has stated that the visual imagery utilized in *Aeon Flux* is done for a purpose. He began the series with a desire to deconstruct typical ideas of heroism.[8] Typical male-centered action texts present a vision of the world that is highly ordered. The hero is depicted as a unique individual, who, encountering an unstable situation (whether it be in the form of injustice, rampant crime, Indians, or evil bureaucrats), establishes order, taming and containing all that is uncontrollable or Other. Linked through filmic techniques to the hero's gaze, viewers look out at a world that is ultimately rendered predictable, safe, and known. Although there are many ideological implications of this structure, action texts work to shore up the boundaries between self and other, life and death, and the known and the unknown, most often through the control of the mind over the body and the punishment and containment of the feminine. In contrast, *Aeon Flux's* female hero, through her "feminine" association with destruction, anarchy, irrationality, and liminality, is used to dissolve the traditional masculinist model of heroism based on the containment of the Other.

Chung, a graduate of the California Institute of the Arts and a former animator at Disney, maintained control over the series as writer and director. Animation for the series was handled by two Asian boutique studios, Mook in Japan, and Gana in Korea. Along with international production facilities, the series utilized individual artists working in diverse locations such as England, San Francisco, and Los Angeles. However, Chung has credited MTV with allowing him the freedom and creative control needed to maintain his vision of the show.[9] Chung has called Aeon Flux a "director's show" in which he and co-director Howard Baker maintained the cohesiveness of the project by piecing together the bits and pieces created around the globe and tying them together. Chung, whose art was influenced by expressionist Egon Schiele, stamped his signature on the visual aspect of the show by instructing the artists on his style, providing character model sheets and creating all storyboards.[10]

Aeon's appearance is sexualized because Chung wanted to represent the "seduction of violence." Aeon's physical appearance and her animated motion are designed to call attention to viewers' visceral attraction to violence.[11] Despite some intervention by MTV's board of standards and practices, which reduced the series' violence and sexuality for the full-length series, the fact that Chung has more on his mind than cheesecake is apparent by his explanation for how he came about with a name for his hero. "'Aeon,' apart from meaning a long period of time, has Gnostic meanings also; the aeons were the intermediaries between the mortals and the divine plane of existence."[12]

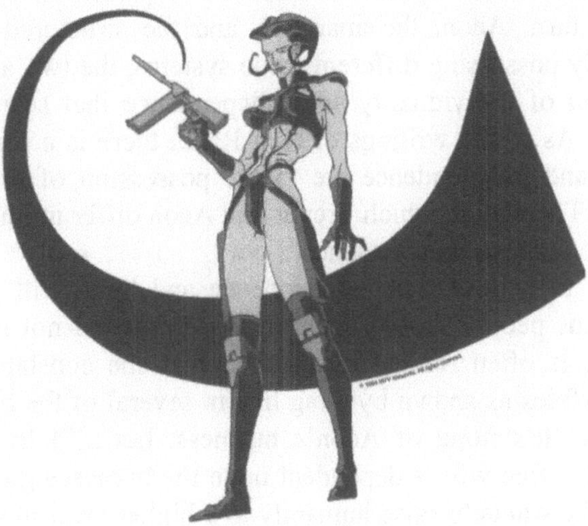

Figure 4. Aeon Flux.
Aeon Flux ©1998 MTV Networks. All Rights Reserved.

Aeon's identity is unclear. Although she appears to be an assassin of some sort, her loyalties remain ambiguous. She may be a double agent working for (or against) her nemesis and sometimes lover, Trevor Goodchild. Goodchild, who is as blonde and Aryan as Aeon is dark, is the head of Bregna, a country with a "structured" society. Like East and West Berlin, Bregna is separated from Aeon's native country, Monica, by a wall. Monica is a "free state" which supports anarchy. The two countries seem to be at war. The ideology of the two countries is never fully articulated, though the ambiguity of Aeon's character and actions suggests that she supports the anarchist views of a Monican. However, there are also indicators that she is actually working for Trevor, whose own goals seem clear: He wants a well-ordered society with highly evolved, independent, and rational humans (under his control, of course). Trevor is a visionary who seems to believe that his plans will benefit the greater good. However, his methods of achieving his goals are highly suspect.

Superficially at least, the series articulates the tensions between the differing philosophies of control versus anarchy, as represented by Bregna and Monica and, in turn, Aeon, the anarchist, and the structured Trevor. Although apparently possessing different value systems, the two actually share a common notion of individuality and independence that seems borrowed from Ayn Rand. As in the writings of Ayn Rand, there is a suggestion that "true freedom" and independence are in the possession of those who are naturally strong. The way in which Trevor and Aeon differ in this philosophy is in the means to achieve independence.

Aeon seems to believe that independence and "free will" will evolve naturally, and thus people should leave things alone and not meddle. (Her stance, however, is often refuted by the fact that she constantly meddles, snoops, and interferes as shown by a tag line in several of the opening credits, which states "It's none of Aeon's business, but....") In comparison, Trevor's desire for free will is dependent upon the harnessing of nature. He often states that he wants to raise humanity to a higher level in which everyone will be free. However, the means he takes to achieve this freedom for others is through fascistic control.

The two characters represent the dualist split of mind versus body and of construction versus destruction. Trevor espouses a desire for rationality and escaping the limits of the mundane flesh. He attempts, through the use of science and machines, to alter the environment as well as to perfect the human body and mind. Trevor builds things: cities, plans, machines, armies, and civilizations. Aeon, on the other hand, represents the "feminine" power of destruction and nature, a Siva-like figure who constantly works to thwart Trevor's plans. She urges Trevor to leave things and people alone, suggesting that natural processes are best. Aeon's name suggests her association with such natural processes. Her name connotes timelessness, while flux is change. Just as Aeon dies only to reappear, her name evokes an endless ebb and flow as well as the cycle of life and death.

The contradictions and similarities between their two positions are a continuing theme in the series. Indeed, notions of independence, individuality, the futility of purpose, and the ever-present use of irony and paradox are the most consistent elements of the show. The early shorts featured a voiceless, characterless (with the exception of her noticeable curiosity) Aeon. Although episodes occasionally focused on Aeon, often she appeared only as a minor or supporting character in films with large casts of equally voiceless characters or as a unifying theme, tying seemingly random scenes together. In these shorts, her identity is thoroughly ambiguous. It is unclear whether she is good or evil or what the purpose of her actions is. In these early incarnations, Aeon always dies, only to reappear in another short. Her frequent death and rebirth suggest, as does her name, timeless, recurring, never-ending change and lead us to ask: Is she just one woman who keeps being reborn, or is she representative of all women or all humans? Although nameless at this point, Trevor also appears and reappears in various roles, such as a soldier who kills Aeon, a fellow agent, and even a naked captive imprisoned in Aeon's kitchen cupboard.

The symbiotic relationship between the characters of Trevor and Aeon reflect a balance between control and anarchy that is demonstrated by both the dialogue and visuals of the half-hour series' title sequence. The dialogue is as follows:

Trevor:	The dream to awaken our world.
Aeon:	You're out of control.
Trevor:	I take control. Whose side are you on?
Aeon:	I take no side.
Trevor:	You're skating the edge.
Aeon:	I *am* the edge.
Trevor:	What you truly want, only I can give.
Aeon:	You can't give it, can't even buy it, and you just *don't get it*.

In this exchange, Trevor presents himself as a visionary and revolutionary. He sees his actions as a form of taking control. Aeon's own character as "the edge" suggests her to be a dangerous, liminal figure with all that liminality suggests: transcendence, corruption, danger, progression, transformation, and change. Moreover, whatever it is that Trevor offers her is nothing that he can control or possess: Aeon is after something intangible, some ephemeral thing that Trevor's rational mind cannot understand.

The first image in the title sequence is typical of how *Aeon Flux* as a whole underscores the relativity of perspective. The first shot is of an eyeball with flies buzzing around it, embedded in some sort of machine. The animation pulls back to reveal that in fact the "eye-ball" is a large statue by which soldiers are marching. The next shot is of Trevor watching the parade from a platform above. The platform pulls backward as the camera abruptly changes angles, dislocating the viewer's sense of perspective.

The following scenes play with images of power and surveillance, setting up an environment of deadly spaces: Trevor walks down a hallway lined from floor to ceiling with TV monitors filled with images of Aeon. The camera zooms in to an extreme close-up of Trevor's eyeball, in which Aeon is reflected. The reflection turns into the fighting figure of Aeon, who leaps to the edge of a tall building. As she runs, she must thread an obstacle course as barriers suddenly appear before her, finally leaping into empty space. Trevor watches as she falls through a series of net-like wires that lash out as if to trap her. As she is about to hit the ground, the wires form into a net that supports her. She places a small, eyeball-like object onto the floor, paralleling the larger eyeball statue in the first shot. The eyeball pops open. It is in fact a bomb, which explodes. The next frame is a close-up of a pool of red fluid with two scorpion-like animals threatening each other. Aeon's shadow is reflected in the pool which splashes onto the tip of Trevor's shoes. The next frame pulls back to reveal Trevor watching Aeon's silhouette in a doorway as the door slams.

The final shot of the title sequence connotes images of corruption and disease. It is an extreme close-up of Aeon's eye with the iris rolled back and a fly buzzing around it. The eyelashes suddenly reach out and trap the fly, and the iris rolls forward. The pupil dilates as the title flashes across the screen and Aeon's theme music plays. The images in the title sequence are typical of the series as a whole. They reflect the main motifs of the series: the use of dangerous, liminal spaces; themes of bodily corruption and contamination; and concerns with relativism, power, and surveillance.

Dangerous Spaces

The confusing, changing, and deadly physical environments created in the title sequence are typical of the series as a whole. In "Ether Drift Theory," Aeon breaks into a "habitat" that Trevor has created. It is a massive, block-like building floating in an ocean of paralytic fluid. The habitat is a "perfectly balanced ecosystem" filled with Trevor's menagerie of strange creatures and mutants. Around the habitat, the bodies of criminals and habitat escapees float imprisoned in the fluid. The habitat itself is riddled with traps and occupied by dangerous creatures, including a huge robot guard, an army of insect-like robots, and killer bees. The habitat is eventually destroyed by an acidic fluid, dropped by Aeon, that rots away the metal. Aeon and the other characters try to escape as it crumbles around them and is flooded by the paralytic fluid.

Monica and Bregna are likewise separated by a border mined with explosives and surrounded by laser towers. Those who attempt to cross the wall risk being paralyzed by stun rays and having their limbs amputated by robots. Similarly, "Purge" begins with Aeon chasing a vicious psychopathic criminal, Bambara, on a train. The train, traveling at high speed through a desert wasteland, is a curious space, housing a wide array of people engaged in strange occupations. For example, as Aeon jumps aboard the train, she breaks a hole into its walls. Despite her presence and the open wall behind him, a passenger crouches on a toilet, unconcernedly defecating. A little boy swings in a swing set that hangs in the gap between two cars, unaware of the track speeding beneath him. Likewise, in the short "Gravity," much of the action occurs as Aeon is falling from a plane to her death.

Just as the train moves over the tracks, the spatial landscape in the show never remains still. Often, landscapes disintegrate, as in "Ether Drift Theory." However, landscapes often change into something else entirely. The environment is full of trap doors, secret tunnels, and moving ramps. In "Purge," for instance, Trevor captures Aeon. Awakening in what appears to be his lab, he turns her chair around and pulls back curtains to reveal that they are in fact on a stage in front of hundreds of people. The stage has trap doors and flying wires controlled by various levers on the stage and amongst the audience. Characters on stage are moved around by these devices, dropping through trap doors, flying in and out on wires, literally displaced by the environment.

Like the weapons of destruction that fill the show, the overriding sense is of an environment designed to trap, not only physically but visually. The animation continually creates the illusion of camera movements. "The camera" is highly unstable, constantly moving, replicating the effect of zooms and pullbacks, which abruptly change the viewer's spatial relation to objects on screen. This creates "trick" perspectives, such as those in the title sequence, in which what first appears to be an eyeball turns out to be a huge statue. In fact, part of the sense of instability the viewer feels is created by the visual tricks that catch out the viewer in her own expectations.

The uncertainty of sight is worth particular attention since visual metaphors are so much in evidence. For example, the title sequence highlights the importance of sight through such images as Aeon's reflection in Trevor's eyes, shadows, and silhouettes, the monitor-lined hallway, Trevor's observation of his troops, and Aeon's eyes. *Aeon Flux* continually plays with the contradictory ideas of surveillance and spying as a means to power, at the same time indicating that appearances are untrustworthy. In the title sequence, for example, we see that an eyeball may in fact be a statue; a dead eye is alive; and reflection becomes the real thing; nothing is as it seems to be on the surface.

In contrast to the narrative films examined by Laura Mulvey, which feature a clear position of enunciation and an obvious filmic subject, it is unclear who the subject of the gaze is in these episodes. Changes in perspectives, which flip between Aeon and Trevor, for example, work to physically dislocate the viewer. For example, in the title sequence, both Aeon's and Trevor's voice-overs are laid over a series of images in which they are not seen. In his first scene, Trevor is surveying his troops and then watching Aeon, but his status of "watcher" is quickly reversed. Throughout the introduction, Aeon's image becomes less visible while Trevor finds himself being watched rather than watching. Aeon's image becomes more and more shadowy: She appears first as a reflection in Trevor's eye, then a shadow, then as a silhouette. Finally, Aeon slams a door as Trevor watches, shutting off his vision. Similarly, the series' bumper featuring an eyeball with the rolled-back iris suggests that she was waiting to ensnare the fly. Though the eye appears to be dead and sightless, it is still waiting and watching covertly. Even though Aeon's physical body disappears, her power of sight is used to remind the viewer of her now invisible presence and suggest that perhaps that she is watching the viewer. The play of power associated with sight is iterated in several episodes and will be discussed in more detail later.

Contaminated/ing Bodies

Just as the physical environment is dangerous and untrustworthy, so is the body. The show juxtaposes images of Aeon's and Trevor's extreme physical strength and agility with images of bodily corruption and contamination. Not least among the bodies is Aeon's own noticeable physique. However, although she at first appears to be a very erotic dominatrix, especially since her character is very sexually active, the visual effect of her body in motion is quite jarring and actually works to counteract the overt sexualization of her character.

Aeon moves in a jagged, insect-like way. Her body, with its black armor, spindly long arms and legs, contrasted to her shorter midsection, enormous breasts, and perky, bobbed, angular hair, give her the physical quality of a spider. Her body is comprised of sharp angles. With the exception of her breasts, which are rather pointed, Aeon has no fat. Angular ribs, shoulder blades, hips, and most noticeably, buttocks, jut out. The animation jerks in such a way that she appears to skitter over the landscape. Her movements are those of an insect: jagged, angular. Her body is drawn in a way that emphasizes the contradictions in her movement, with, for example, her head turning in one direction as her legs go in another. The movements do not emphasize her grace; quite the opposite, she usually lands in an animal-like crouch. In addition, her body is often killed, crumbling into grotesque angles as it lies crushed and bleeding.

While some viewers undoubtedly find Aeon's body titillating, there is a certain quality of grotesqueness about it. In addition to her sheer angularity and insect-like movements, she is often seen from angles that emphasize her buttocks and genital region and inner thighs. Certainly, her chastity belt only emphasizes the pubic triangle it is "supposed" to hide. Indeed, her anatomy is not only revealed, it is overemphasized: We can see into the joints of her crotch. Like the opening shot of "Gravity," which features a mouth's point of view of a tongue entering it, her appearance is a bit too intimate, too detailed to be erotic. An essential element of the erotic is the element of mystery. Aeon's body is not erotic, because the very volume of naked butt shots, breast shots, and crotch shots immunizes the viewer against their shock value. By displaying her body and its parts so obviously, some of the power of the image is removed. Her partial nudity becomes interesting only because it signals the vulnerability of her flesh: a flesh that dies in contrast to her sheer physicality and physical power.

This contrast between physical strength and vulnerability is central in *Aeon Flux* and points to the series' motif of bodily contamination. Contamination may occur through both natural or technological means. Things associated with the body such as eating, sex, and death are depicted in a way that reveals a horrified fascination with the permeability of the body. Bodies are under constant threat of contamination, and to resist contamination is to maintain one's control and individuality.

Scafandra, another female spy, has hands to replace her missing feet. Her tongue has a hidden slot from which darts and wires spring forth. Grappling hooks and wires also spring from compartments in her heels and hands. In "Thanatophobia," the character Sybil has her back broken during an attempt to cross the Monican border. Trevor and his scientists remove her spinal cord and insert a hinge, which keeps her back straight as long as she locks it with a fluid-filled key. When the key is removed, Sybil can contort her body like a snake.

Clavius has his stomach turned into a tunnel in "Utopia or Deuteronopia." His belly is held together with series of metallic locks; he is kept alive by being plugged into an outlet in the wall. In "Purge," an artificial "consciousness" made of metal is inserted into various characters through their navels. The navel itself may be pulled out and looks like the twisted knot of a balloon. The belly can be opened simply by untwisting the knot. In "The Demiurge," Trevor removes a baby alien, called the Demiurge, from the impregnated body of Nadir Zenith. Poking a compass-like object into Nadir's flesh, the flesh, like a mechanized trap door, slides back into a perfectly round circle. There is no blood, just a glowing open cavity through which the baby Demiurge floats.

The body is particularly subject to contamination from natural means. The body as a source of decay is depicted in the very first shot of the title sequence which features flies buzzing around the eyeball. The flies symbolize corruption, both through their association with death and the processes of decomposition, as well as the devil. Death is similarly evoked in the exploding eyeball, which appears to also have been some kind of egg. The squirming scorpions that are revealed by the explosion had been a secret, hidden within the depths of the eye. The imagery of hidden corruption is reinforced by the plots of numerous episodes.

Infectious viruses are also a recurring motif. Bargeld in "Ether Drift Theory" is infected and finally killed by a virus he himself has created, which also infects his girlfriend, Lindsa. Throughout the episode, scenes are intercut with shots of the microscopic view of the virus as it grows. In "Isthmus Crypticus," a winged alien is killed by a single insect bite that instantaneously rots her skin. In "End Sinister," Trevor releases a virus into the environment. The virus is shown spreading over the bodies of aliens who arrive on the planet in their spaceship. Like crescent shaped moths, the virus dots the bodies of the aliens and later appears to have taken on a life of its own, hovering in the air. Viruses spread instantaneously and inexorably.

The threat of bodily contamination is often linked to sex, tying sex to horror, pain, and victimization. In "Purge," the body is literally invaded by a metal consciousness, which thrusts itself into the navel of its victims as they arch their backs and howl. It is hard to tell if characters are yelling in pain or in orgasm. Similarly, when characters are bitten by bugs, cut, etc., their howls of pain are often orgasmic in tone. Technologically altered bodies, while grotesque, are often sexualized. For example, Trevor brings Sybil, the girl with the hinge in her back, to orgasm by fiddling around with a surgical instrument in the open hole in her back. Meanwhile, the inserted hinge leaks fluid. Similarly, the aliens in "End Sinister" have sex by exchanging eyeballs with each other and Trevor, suggesting a permeability and mutability of the body. The aliens carry spare eyes with them in a sack. Aeon, investigating the alien space ship, pokes her finger into the eye sack and pulls it out dripping with fluid.

Even without the use of technology, sex is threatening. For example, the short "Gravity" opens with a mouth's point of view shot of a tongue inside it. The tongue rolls around and hides a document in a secret compartment in a tooth. The camera pulls back to show Aeon and Trevor kissing. When characters kiss, the sight is usually accompanied by a slurping sound. Characters' tongues often in and out of their mouths like long wet snakes. Often tongues are inserted into ears and pumped in and out like pistons. The visual effect is one in which a "romanticized" notion of sexuality is denied. Body parts such as tongues are visually depicted as grotesque. In "Reraizure," Rorty, one of the many amputee characters that fill the show, has a stump for a leg. Aeon strokes it like a phallus. The visuals and sound effects evoke a repulsed child's view of sex. Importantly, the body, structured as permeable, leaking, and wet, evokes the female.

Not only is the female body grotesque, it is also dangerous. Aeon Flux is physically dangerous to her foe, but her sexuality is even more dangerous. The novel *The Herodotus File*, for example, features a sequence that blatantly plays upon the male horror of female sexuality. Trevor has decided to assassinate Aeon through the means of bombs inserted into his sperm, which he plans to ejaculate into her. After Aeon and Trevor have sex, the next sequence of frames shows an interior shot of Aeon's vagina as the explosive-laden sperm travel towards her uterus. Lying in wait for the bomb, however, is a multi-fanged robotic device, which eats the sperm. Aware that the bombs have been aborted, Trevor pulls out of Aeon in dismay at she laughs at him and asks: "What's the matter, Mr. Chairman. Is it something I did?"[13] The castrating threat of the vagina dentatis is evoked along with the threat of female laughter and sexual ridicule. Additionally, such sequences play with the idea of bodies being infused with materials from the external world, contaminants that threaten bodily destruction.

Two major issues become clear through the theme of contamination. First, the series continually portrays the destruction of plans by the seemingly small and random. Second is the linking of contamination and destruction to Aeon herself. It is no coincidence that it is Trevor, the intellectual man of science, who is constantly trying to construct order and that it is Aeon who destroys it. As Trevor yells at her in "Ether Drift Theory," "Your vapid heroics are contaminating my properly balanced eco-system." Similarly, in "Isthmus Crypticus," Aeon, seemingly out of jealousy, but perhaps altruism, breaks into a special habitat designed by Trevor to imprison the angel-like winged aliens. However, the habitat, while a prison, also keeps them safe. Aeon blasts a hole in the wall allowing the angels to go to each other and apparently regain their power. But in creating the opening, she also carries in the insect that kills the female angel and later kills Aeon's friend, Una. Similarly, Aeon releases the virus that kills off humanity in "End Sinister."

Indeed, Aeon's very presence can be disruptive. For example, in "Reraizure," Aeon has broken into Trevor's fortress to steal photos of them making love. On her way out, she encounters Rorty's girlfriend, Muriel. She tells the girl she means her no harm and "let's just pass like ships in the night," but Muriel stabs her as she passes by. Aeon's blood falls to the floor, wetting the grate on which Muriel stands. Muriel slips and falls through the grate to her death.

The ambiguity of Aeon's character emerges against this setting of corruption and contamination. Paradoxically, Aeon is depicted as both the disease and the cure as in, for example, the final shot of the title sequence, in which her eyeball captures the fly. This shot is used as a bumper throughout the series and has an iconic function, signifying Aeon. The shot creates a paradox between hunter and hunted. Initially, it seems like the fly is preying upon a corpse but is itself ironically made into a corpse; Aeon has been preying upon the fly. The comparison to a Venus flytrap is quite obvious. Again, is this an act of benevolence or cruelty? The fly symbolizes death and decay, and she has caught and trapped it—and yet it is also just a fly, innocently going about its business. Is she the virus or the vaccine? As these scenes make clear, the series presents truth as hidden or at the very least, dependent upon one's viewpoint. Everything is relative.

Aeon Flux has a gleeful sense of the paradoxical, ironic relativism linking life and death. Small and random events lead to destruction. It is no coincidence that viruses are a recurring image. The virus is a living being, unnoticed and invisible, inconsequential except for the devastating effect it has on life. In addition to viruses, characters are often killed by bug bites. In "Ether Drift Theory," Trevor recruits Agent Sixteen, a female martial artist with four arms, to capture Aeon. Agent Sixteen shows off her martial arts ability in a complicated demonstration. Only seconds later she steps through a doorway, is attacked by a swarm of bees, and immediately drops dead. Small, seemingly inconsequential details upset well-laid plans and destroy individuals.

Similarly, "War," is a short that follows a progression of characters through a battle, whose purposes and sides are unclear. It starts with Aeon, who climbs a plateau after killing hundreds of Breen soldiers only to be killed in the first few seconds by the one surviving Breen soldier, an early incarnation of Trevor. We then follow the Trevor-figure charging through an army of Monican soldiers and breaking into a Monican stronghold, where he is in turn killed by a Monican guard. The Monican then single-handedly fights off an attack by a Breen warplane. One female Breen soldier manages to break into the Monican fortress and meets up with another male Breen soldier. They kiss and continue on their mission, as the short ends. Meanwhile, the action sequences are interspersed with shots of oil dripping from a pipe in the Monican fortress to puddle on the floor. As the drops continue to flow, the puddle spreads. This is a rather obvious visual metaphor for war, connoting spreading pools of blood as well as the vast numbers of individuals whose individual stories progress or end even as they connect to the unknown stories of others. In this way everything is linked and individuality is denied. Aeon and Trevor and thousands of others can die, but life continues.

Another way in which the series plays with the concept of relativism is by denying the power of action. Individuals are constantly hoist by their own petards. For instance, in "Ether Drift Theory," Aeon and Lindsa break into Trevor's "habitat" to rescue Lindsa's lover, Bargeld, a scientist. Bargeld is ill and eventually dies from a virus that he himself has created. Similarly, Aeon, who enters the habitat with only the best intentions, ultimately ends up destroying herself. Entering into a restaurant that was designed to feed the inmates, she steals their food. While eating, she accidentally knocks over a glass of liquid that combines with an egg that has broken on the floor. The combination of material is deadly. Turning into a green fluid, it rots away the metal of the habitat and spreads. The building disintegrates and traps Aeon in the paralytic fluid. Meanwhile, despite Aeon's best intentions to help Lindsa, Lindsa turns against her when she thinks she sees Aeon having sex with Trevor. Trevor has, in fact, wrestled Aeon to the ground. Actions are interpreted and misinterpreted, leading to the exact opposite of the desired result. The series ultimately suggests that any action that one takes has consequence, whether intended or unintended. Thus, any action, even non-action, creates effects. Action is thus political. This discloses the relativism of actions. Which actions are wrong? Which actions are right?

A case in point is "End Sinister." In this episode a group of dying aliens are monitoring the planet from a space ship. The sole survivor exits the ship in a stasis chamber that lands on earth. Meanwhile, Trevor plans to release a virus with the means of a ray from a satellite. The virus will kill the majority of the human race. Typically, he sees this as a humanitarian act that will strengthen the human race by helping it evolve. As he says:

> Strength is not an absolute value. To be strong is to evolve. Mutability is strength. Evolution is a string of victories by the strong over the meek. The meek shall not inherit the earth. What is strength when a 200-pound man can be killed by a ray emanating from a satellite in space? Strength is within. Case in point. I make two men sick by exposing them to the same ray. One dies horribly, with pain. The other buries him without remorse. He is the winner. Who will the other winners be? Those who will be left once I use the ray. We have already evolved so much, our actions

> would be incomprehensible to a human from a thousand years ago. It's the evolution revolution. Tomorrow we speed up the process. All sectors. All species. Only the strong shall evolve.

Aeon breaks into Trevor's bedroom to stop him. Telling Trevor she won't let him destroy most of the world, she steals the detonation device. They argue. Trevor tells Aeon not to worry because the ray will only affect the weak but will make stronger those who are already strong "like you and me." Their strength is at least suggested by the fact that Trevor believes mutability is strength, and the two of them are certainly mutable, changeable. Aeon angrily knocks him unconscious, saying, "nobody can make me be anything." Her very mutability is something that resists change through manipulation.

Aeon escapes into a forest and hides the device; there she encounters the alien in its escape pod. She is startled to see the alien has no mouth, nose, or ears, and when it takes off its cloak, it has no midsection, just two long legs that reach up to its shoulder blades. She warns the alien that it is in danger and tries to help it escape. Eventually, the alien is captured by Trevor. Trevor immediately becomes intrigued and fascinated by the alien, who mysteriously carries strains of his virus.

Trevor is infatuated with the "physical perfection" of the alien. The alien has moved past all weakness of the flesh. It communicates telepathically and has no need to eat or defecate. The alien's mind and intellect have evolved to a high level. As Trevor says "She lives in the mind. Her feelings are real, more intense than our own. Her language of pictures is both poetic and more precise. She *needs* no one else but she *wants* someone else." The alien has the freedom of *choice*.

Both Trevor and Aeon develop a relationship with the alien, who, because it has no bodily orifices, mates by exchanging eyeballs. Aeon rejects the alien's pass, telling it, "There are some things even *I* won't do." Trevor, however, seeks to become like the alien whom he sees as representative of what human beings might become. He exchanges eyeballs with the alien and experiences a change of heart. He decides not to release the virus. Desiring to evolve into a higher life form, he plans to go with the alien to its home planet. Aeon discovers the spaceship and sneaks into Trevor's bedroom to confront him.

Moving in and out of the shadows of the bedroom, lit by the light from an open doorway, Aeon accuses Trevor of letting the alien use him for its own ends and asks him how he can stand being controlled. In a typical Trevor-Aeon exchange, he replies "I would rather be under control than *out* of control like you." He then tells her he was ashamed of his plan to release the virus. Aeon is horrified. She tells him he is running away and forgetting what it means to be human. Obviously upset, she warns him that he won't be able to make the trip. He turns to face her, for the first time revealing that he has exchanged an eye with the alien. "But I have made it," he replies. "So I see," says Aeon, and disappears out of the lit doorway. Trevor yells after her, "No, you don't see. You need to see. We *all* need to see. And we will…".

Trevor and Aeon are both connected to metaphors of sight and light as forms of knowledge and truth. They both claim to "see," though what they "see" is diametrically opposed. Both share the notion of a higher individuality. However, Trevor seeks to claim individuality and control through the shedding of bodily desire, through orderliness. What he sees as Aeon's being "out of control" is her version of independence, in which she is dependent on no one.

In this way, Trevor and Aeon represent the mind-body split. Trevor, the scientist, fools around in his lab, investigating and probing the alien, creating new life, viruses that are, paradoxically, designed to destroy life. Like the intellectual, rational mind, he seeks to escape from the earthly mundane inconveniences of the polluted flesh. Aeon represents the irrational, lived body, who rejects the control of aliens and others.

This fact is underscored by the final moments of "End Sinister." After the alien and Trevor leave in the ship, Aeon says that Trevor will be back and she prepares to wait for him in the stasis chamber. When she wakes in the future, she finds a city filled with aliens. Trevor is there nursing an alien baby. Horrified, Aeon encounters the original alien and yells at it: "Is this what you wanted? Now I'm the only one." She races to find the trigger for the virus ray that she had buried centuries before. As she runs, she fights off images that the alien sends her of being joined, of becoming one with them, and is almost slowed to a halt by an image that Trevor sends her of them kissing. Aeon is terrified of the promise of unity offered by these images. She releases the ray, and the aliens immediately become ill and die.

Trevor joins her in the forest and tells her it is ironic that it all started when she stole the ray because she didn't want him to kill off half the human race, yet she has just killed off the entire population. The "aliens" are, in fact, human beings who have evolved to their present unrecognizable state, their actions "incomprehensible" to Trevor and Aeon, the humans from centuries before. She loads Trevor and herself into the stasis machine. Meanwhile the surviving aliens climb into the spaceship, heading back in time to where the episode began. The final scene shows an ancient stasis chamber buried in rubble with its light beeping.

This episode shows the chaotic effect of any choice made by *anyone.* Aeon struggles to maintain her independence against visions of unity offered by the aliens and Trevor, and yet she is dismayed to be "the only one" left.

Trevor, who seeks to be superior, finds the peak of evolution to be unity with others. Finally, Aeon actually commits the very act that she set out to prevent: the destruction of humanity. Like the relativism of a tack killing a seemingly invulnerable warrior, end results, which are outside of human control or knowledge, take precedence over human desire and will.

Sight and Power

In the world of *Aeon Flux*, knowledge is always false because the "truth" is impossible to know. "Utopia or Deuteronopia" exemplifies the way the series as a whole critiques rationalism, the idea that the truth is discernable and quantifiable, through the metaphors of sight and Aeon and Trevor's relationship. The episode begins with a synopsis telling us that President Clavius of Bregna has been missing. Meanwhile, Trevor has seized power. The action opens with a voice-over from Trevor, which is actually a pretty straightforward summary of the plot:

> There can be no justice where there is no truth. What is truth? Tell me if you know. And I will not believe you. Things are never what they seem. Clean gloves hide dirty hands. And mine are dirtier than most.

Trevor has kidnapped Clavius and jailed his supporters. Meanwhile, he is dealing with public relations problems. He is interviewed by a reporter, whose floating camera, shaped like an eyeball, broadcasts his image to massive television screens placed around Bregna. His interview is watched by crowds of people around the city as Trevor outlines his new plan of an "open society." He informs the crowd that only an open society can be a just society and strips himself of clothes. He tells the audience that he doesn't "indulge in feelings" in regard to his detractors and that "Nothing is sacred, Nothing is secret."

Meanwhile Aeon has broken into Trevor's palace, informing security cameras that she is there to kill Trevor. She asks them if they believe her. She meets up with Guildemer, a subaltern who is loyal to Clavius and suspects Trevor of foul play. Guildemer has found a scrap of cloth dropped by Trevor on which code numbers are written.

Trevor goes to bed where he is filmed by a video camera that is being monitored by Guildemer. Trevor disrupts the broadcast, places a dummy in his bed, and climbs down a secret tunnel beneath his bed that leads to a chamber. He puts on a harness that shakes him back and forth at a tremendous speed. Vibrating, he is now able to see another vibrating figure in the corner, Clavius, who is hooked into a similar harness. Clavius has several metallic objects imbedded into his skin and appears to be kept alive by means of an electric wire plugged into a socket. Unconscious, he is naked and obese, his grotesque appearance enhanced by the man-made additions, including a padlock and what appear to be several large holes drilled into his stomach.

Trevor lovingly shaves Clavius, telling him that he is performing this service in gratitude for what Clavius has given him. He then pulls out a key, opens the lock in Clavius' stomach, and steps into his belly, which becomes a long, red tunnel. Crawling along the tunnel, he comments that Clavius has given him the way to Aeon Flux: "true freedom. Freedom from prying eyes and expectation. A private place. A place of our own." He enters into a red-curtained chamber with a bed, a bouquet of roses on the floor and a dummy in a long black dress.

The tunnel and hidden chamber are a conundrum since they are too large to be inside of Clavius. They are red and at times appear to be his insides, with supporting beams that look like ribs. However, the vaginal reference is unmistakable, while the red-curtained room, a place of peace and security, is obviously a womb. It is soft, surrounded by flowing red curtains, graceful, sweeping, curving lines, rather than the hard sharp angles of the exterior industrial world. This secret realm created by Trevor exceeds the boundaries of its container. The interior, secret, hidden world is larger than the outside.

Later, Trevor watches a wall full of security monitors that broadcast "secret" behaviors of those being spied upon, among them a sadomasochist couple. He muses on the benefits of his new "open" society. He says that openness means that there is nothing to lose but "shame, inhibition, and perversion. All the good things in life." Shame and perversion will be done away with, but they are, paradoxically, good things. Trevor himself wants to keep secrets such as the hidden room in Clavius' stomach because they offer freedom. Secrets—the unknowable, the unseen—offer freedom.

Through the security monitors, Trevor spies on Aeon Flux meeting with Guildemer, plotting to destroy him. Trevor laughs and then talks to the screen, urging Aeon to look at him. As if she has heard him, Aeon turns and looks up at the screen. She does a "now you see it, now you don't" magician's wave and pulls out the piece of cloth. Trevor freaks out and attempts to capture her. She knocks him out and ties him up behind a screen in an empty room. Gagged and bound, he is forced to watch through a slit in the screen. Aeon, now dressed in a red leather thong outfit, reclines on a table, facing him, with her legs apart. The door to the chamber opens, and a sweating Trevor is forced to watch as a nameless guard sucks Aeon's foot.

Meanwhile, Guildemer has found Clavius and discovered the key to his stomach. He does not quite know what to do with it. He interrupts Aeon with the guard. Nervously glancing at the screen, he tells Aeon that he feels he is being watched and discusses his plans to get rid of Trevor. They begin to make love, but his attempts are stopped by Aeon's chastity belt. Guildemer leaves after giving the key to Aeon. Aeon goes behind the screen to talk to Trevor, but he has burned through his bonds with his cigarette lighter and escaped.

Cut to Trevor rushing to Clavius' cell. The door in his belly is already open. He crawls down the tunnel and peers through the curtains into the red-curtained room. Aeon is already there, fingering the dress. Aeon mocks him for spending his energy on such trivial matters. She steps carelessly on the bouquet of roses. Trevor is hurt and tells her she should kill him. Instead, she kisses him, saying, "You think you know what I'm doing; so obviously, you don't." She turns off the vibrating harness, which knocks Trevor unconscious. When he awakens, he runs down the tunnel, which has suddenly become much longer. It opens out into the middle of a forest; the tunnel appears to stretch behind him for miles. The secret, hidden place, larger than the outside, has grown larger and is now a mystery even to Trevor, reflecting the mystery of Aeon, whose own hidden internal world is unknown to Trevor.

Meanwhile, Guildemer awakens Clavius only to discover that he is insane. He unplugs him, and Clavius dies just as Aeon finds them. Trevor, spying on her through a monitor, thinks that she has killed Clavius. In the middle of an interview with a—now naked—reporter asking him about the whereabouts of Clavius, he leads the army and reporters after Aeon. Trevor finds Aeon as she loads a body tied up in a sack into a truck. He tackles her to the ground, crowing triumphantly that he has her now, and she is powerless. He unwraps the sack to disclose a bound and gagged Guildemer. Guildemer, sounding insane, tells the people that Clavius was mad and corrupt. He holds out the key to Clavius' stomach to prove it. Aeon takes the key and uses it to open her chastity belt. Many of the assembled crowd pull similar keys from their pockets. They laugh at Guildemer and imprison him. Aeon throws Trevor the key to her chastity belt, kicks a wall behind her which explodes, and escapes. Later that night, Trevor gazes out of his bedroom window while wearing his vibrating belt. Aeon appears under his window, also wearing a belt. She blows him a kiss and then disappears.

This synopsis reveals how the animators play with the notions of sight and power. The characters are all voyeurs, but rather than giving them power through what they see, their knowledge is unreliable. As Trevor says "Clean gloves hide dirty hands." Eyes are often betrayed. In "Utopia or Deuteronopia," spying on Aeon through his monitor, Trevor mistakenly thinks that Aeon has killed Clavius. "Gravity" features Aeon falling from a plane as she spies upon a truck speeding through the desert. She is so preoccupied with watching, she seems unaware she is falling to her death. In one untitled short, Aeon breaks into a house. She goes into a monitor room that shows the entrance. However, there is a glitch in the monitoring system. Although she repeatedly checks the monitor, she is unaware that Trevor has snuck in behind her. He ends up killing her, as the monitor, which is running on delay, shows his entrance.

Although Aeon is at times fooled by sight, it is Trevor's sight that is perhaps less reliable. His habit of surveillance is turned against him, as in the foot sucking scene in "Utopia or Deuteronopia." It is important that this episode ends with Aeon watching Trevor and maintaining her mystery and her power by disappearing from his view. Such moments convey a sense that mystery and the unseen are more powerful than sight. In fact, "Utopia or Deuteronopia" makes it quite clear that Trevor is in fact trying to preserve mystery and secrets for himself. He acknowledges that secrecy leads to "all the good things" in life. One of those good things might be the secret of Aeon Flux. He builds a hidden room where they can be together without the prying eyes of others. Aeon is thus equated with mystery and secrets, which are also sexual. The tunnel and room appear to be similar to the female sex organs. The "secret" place is the core of a woman's body, and in this case the woman is Aeon. He is building a temple to the mystery of her womanliness. When she rejects him, the tunnel is transformed. It becomes huge, extending for miles, finally exiting into the middle of a forest. Trevor finds himself lost in familiar ground: The tunnel he created has changed and become strange to him. It is a path that leads him nowhere. He is overwhelmed by her mystery. Trevor's spying and his attempts to capture Aeon echo his desire to know the truth. However, Aeon cannot be contained by any rational means.

As this episode reveals, Aeon and Trevor's relationship addresses questions of love, possession, and power. It seems that Aeon and Trevor do love each other. However, they can never be together, because they each, particularly Aeon, need to be independent and uncontrolled and thus have to remain apart. This is particularly evident in "A Last Time for Everything," in which Aeon learns of Trevor's scientific breakthrough in human cloning. Along with another spy, Scafandra, she crosses the border. Scafandra is captured. Aeon goes to her rescue, setting her free just as Trevor steals her bodily fluid. Aeon and Trevor argue over the morality of what he is doing as Aeon allows him to steal a sample of her blood. Later he clones her. The clone takes a bath and swims down a secret tunnel at the bottom of the bath. She meets up with the real Aeon. The two exchange places, with the real Aeon swimming off and reminding the clone to "kill you" at the end of three days when their plan is complete.

Aeon returns to Trevor's bedroom. He orders her into bed and tells her that nobody is there to see what they do. Accordingly, Aeon becomes complacent, eager, and sexually obliging. Turning and twisting into various positions, she asks him how he likes it. Trevor is repulsed by her acquiescence. He tells her to leave and goes off to sulk in his bedroom, surrounded by dozens of Aeon clones. Aeon comes after him and they have sex. The next morning, Aeon looks out the window and wonders where "she is." Trevor suddenly realizes that the woman in front of him is "the *real* Aeon."

But the course of true love never runs smoothly. Aeon is enraged when she learns that Trevor has ordered his men to kill the clone. She angrily tells him the clone is the only reason she is able to stay with him. Like any dedicated career woman, she will only take time to be with her man as long as the job is being done. The cloning allows her to have two lives, that of the action heroine and that of the lover. More importantly, the Aeon clone allows Aeon to be "out there somewhere," to retain her separate identity and her mystery. Aeon and the clone are eventually forced to cross the border together, where Aeon is shot and killed. The Aeon clone runs into the sunset as Trevor orders his men to let her go.

Just as Trevor and Aeon can never be together, other lovers remain separated. In "Ether Drift Theory," Lindsa and Bargeld's affair is doomed. In "Isthmus Crypticus," a male and female alien, who appear to be angels with rainbow-colored wings and piercing, angelic voices, are imprisoned by Trevor, who is in love with the female alien. Aeon frees the female, only to have it die from an insect bite. Meanwhile, the male is set free by Aeon's assistant, Una, who is carried away by the alien to lofty beautiful heights, only to be bitten by an insect whose bite rots her skin. In "Thanatophobia" two Breen lovers seek to cross the border to Monica to be free. When he succeeds and she fails, their relationship changes. The man is revealed to be unfaithful and uncaring of his lover, who is eventually trapped in her attempt to cross the border and has her limbs amputated. Even when lovers are united physically, they are parted spiritually. In "The Demiurge," Nadir Zenith and Celia are both killed and then brought back to life by the Demiurge. The two have a happy love ending only because Celia lies to Nadir about her knowledge of the Demiurge: Lies and fiction keep relationships together; true unity is impossible.

Just as lovers can never be truthful and faithful to each other and be together, the series articulates a constant tension between unity and independence as, for example, in "End Sinister" and "The Demiurge." In both episodes, Trevor is drawn towards unity. Preparing to deliver a "baby" Demiurge that has been planted inside the body of a Monican soldier, Trevor shaves himself, musing:

> Shedding our animal nature, we prepare a vessel for the advent of merging wills. We become like infants, primordial and innocent. The return to the womb. The rebirth. What the Demiurge offers is the opportunity to be made whole. To be one with the eternity residing within us.

Aeon, of course, resists the Demiurge just as she resists the promise of unity with others offered by the aliens and Trevor in "End Sinister." Similarly, in "Purge," Trevor inserts an artificial conscience into a criminal and psychopath, Bambara. The vicious man is turned gentle and loving, cutting off his arm and donating it to an armless little boy. Meanwhile, Aeon joins a group of female insurgents who are opposed to the notion of "false consciousness"—the pun is obvious. They steal the artificial consciences from people's bodies, inserting them into their own bodies to later be removed and destroyed. Aeon traps Bambara and removes his conscience. However, she is unable to make herself insert his conscience, struggling against its attempts to enter her. Later, when Aeon is captured by Trevor, she awakens in his studio and wonders if he has removed her free will. Aeon escapes from Trevor when Bambara, enraged that the implanted conscience made him give away his arm, interrupts them. Leaving the building, Aeon is disturbed by sight of the conscience in a glass container, ominously banging against the walls in its struggle to get out. The sense of peace the conscience offers threatens her.

While Trevor seeks to become whole through whatever means possible, Aeon resists togetherness or completeness. It is left to Trevor the visionary, the representative of intellect, to yearn for more. Aeon is content in the here and now, with what she has, in the lived moment. Aeon is a foil to Trevor, who represents control and the mind. Aeon, as a female, is closely connected to anarchy, mystery, and bodily processes such as sex and decay. Aeon's fear of unity is perhaps due to her unwillingness to be happy or tamed. However, it is as equally likely that Aeon wants to be as she is because she feels she is complete in and of herself. The mind is able to be aware of loss and absence, while the body just *is*. In this way, the two represent conflicting but connected philosophies. Aeon reflects the notion of imminence, of being in the moment. Trevor represents the aesthetic, the point of the reflection of the mind on what has already passed, and a desire to return to it. Importantly, Trevor's attempts lead to failure: Aeon, even after death, remains Aeon.

Life in Death

There are some interesting comparisons to make between traditional action narratives and *Aeon Flux*. In *West of Everything*, Tompkins writes that the tension between community and individuality is expressed through the hero's control of his body. The hero seeks to escape into the wilderness away from towns and society, which are equated with the feminine. By sealing himself against human interaction, as with silence, the hero's body is impervious to penetration and the feminine. Likewise, the suffering his body endures demonstrates his masculinity as well as the strength of his mind and his moral superiority.

The notion that the (rational, pure, male) mind must control the (irrational, dirty, feminine) body is not unique to action/adventure narrative. Western medicine has similarly fought to control the excesses of the female body, externally and internally. Foucault's writings on the medical gaze present a world in which a disciplinary gaze constructs the female body as pathological and excessive.[14] The gaze breaks the body into pieces and then names these pieces as diseased, containing and controlling the forces that threaten. Susan Bordo has similarly argued that the disciplined body becomes plastic, with its individual pieces sculpted by the will of the mind. Ultimately, order and meaning are established by the self-regulation of the body. Bodily control is a way to shore up the body against the "feminine" processes of decay.[15]

The hero's body replicates what Kathleen Rowe has called the "classical body," which "consistent with the ideology of the bourgeois individual," shores up the boundaries between itself and the outside world. In contrast, she observes that the "grotesque" body breaks down these boundaries.[16] It is important to note that just as the control of the body, the control of relations to others, is a defense against the feminine, it is also a repression of death. The grotesque body, on the other hand, articulates a contrasting view towards the nature of death.

The connection between bodily control and conceptions of mortality is described by Rowe in her discussion of carnival and the grotesque. As she observes, the carnival is a "struggle over social and political power, the triumph of the natural over the abstract and the communal over the individual."[17] Moreover:

> This material principle of life and the values of becoming and relativity, which Bakhtin sees in the dialogic play of language, are epitomized in an aesthetics of the *grotesque*. The grotesque exaggerates incompleteness, process, and change, maintaining a kind of moral neutrality or *ambivalence* toward time and death. When life is understood from a collective rather than individual perspective, time becomes the necessary link in the process of communal growth and renewal, and death the "other side" of birth rather than a source of terror which irrevocably separates the individual from all others. For that reason, dismemberment, horror, death and taboo are essential elements of the grotesque. They are reminders of the totality of life, possessing a demonic power to unsettle all that would conceal the democratic materiality of the body.[18]

When viewed from a perspective that embraces collectivity, death is nothing but a part of change and process as, even though an individual dies, the collective continues. In this view, the fear of the loss of individuality is contrasted to a desire for unity. One view suggests the end, the other, continuing process.

The landscape in *Aeon Flux* reflects this process. It is constantly changing and unstable. There is no sense of reference for where the characters are in space. Although the narrative describes a wall separating two distinct realms, the wall appears different in different locations. Both Bregna and Monica look the same, with their rubble-strewn, industrialized space. The land contains pitfalls and traps such as laser towers; it crumbles and disintegrates; it is constantly shifting, a model of change, the processes of decay *as well as* growth and life.

Like the Western hero, landscape is used not only to set up an opposition or a problem for Aeon but to define her. Just as the series' landscape represents the notion of boundaries or liminality, Aeon is also a liminal figure. *Aeon Flux* appears to me to achieve what the Western hero aspires to be: unself-consciously, unreflectively *one* with her surroundings. Yet Aeon is also endangered in this world; she dies, often from a result of the landscape, for example, falling to her death from a high rise. And though she too is reborn, within the individual episode, it is the vision of the landscape that remains to outline her broken body. Yet, as is seen by her name, the series emphasizes continuance. Importantly, Aeon is the one character who seems to be able to pass freely back and forth through the border between Bregna and Monica. She dies, but is reborn. She may be an individual woman or merely representative of Everywoman. Either way, life will continue, even though the individual Aeon may die.

Thus, although it is Trevor who constantly wants unity, while Aeon is the lone individual, their actual actions reflect the opposite. *Aeon Flux* presents a grotesque bodily aesthetic, while Trevor's actions work to underscore his individuality and separation. The natural (Aeon/Chaos) continually triumphs over the abstract (Trevor/intentions). Death is ever present, reflected by visceral, disturbing images of death, mutilated and penetrated bodies. The "democratic materiality" of life is shown in the way that the tiniest of viruses can kill armies. As noted previously, the imagery reveals a fascination with boundaries and borderlands and repeatedly plays on themes of "death-in-life and life-in-death," as for example, its use of infections, and its relativistic view towards life and death.

Many images in *Aeon Flux* may be read as misogynistic. She is presented as a dominatrix, on the one hand, while there is also a horrid fascination with bodily, leaking fluids which references a horror of the feminine.

However, despite, or perhaps because of, this sense of horror surrounding her body, *Aeon Flux* creates an image of an extremely powerful woman. Moreover, the more "collective" view towards life that is expressed by the grotesque body is a dramatic break from traditional action texts.

However, unlike the majority of action texts, the structure of *Aeon Flux* denies viewer interpellation into any particular subject position. The images all create an alienating effect for the viewer, an objective stance, rather than urging the viewer into a position of identification with the heroine. This actually works to create a critical view towards the action hero; the series confronts heroic myths of self-transcendence and mocks heroism, the triumph of the will over the body, individuality, and immortality.

Conclusion

Aeon and Trevor's fascination with spying comes across, finally, as a joke, because their sight does not seem to help them. Just as mighty soldiers are cut down by another soldier and people are killed by viruses or tacks in their shoes, events are finally uncontrollable. What people think they know is not what they know. No truth is known. Chaos, governed by paradox, is the only rule. The beautiful Aeon is at times grotesque. Her physical expertise often does not save her from, but leads to, her death. Whatever her motives are, she brings destruction to herself and others. Trevor's plans to control himself and others never work. The only constant rule in the world of *Aeon Flux* is that both Trevor and Aeon do what they (think) they want. However, their desires often appear nonsensical, destructive, paradoxical, or futile attempts doomed to failure. In short, despite their pride in their free will, they have no control over, or freedom from, the workings of chaos.

Moreover, Aeon's female body is essential, due to the balance set up between Trevor and Aeon. The two must appear as opposites in order to set up the line of paradoxes. Of course, the show plays upon stereotypical notions of the woman as destructive, trouble-making, while the man is rational and concerned with the mind. Her femaleness and "chaotic" quality are enhanced by the equation of the female to the body and the body to the vulnerable female.

The series demonstrates a great distrust of the human body. There are repeated images of tunnels, of bodily permeability (infected by viruses, false consciousness), leaking body parts, oozing liquids. Moreover, the images reveal a fear of the reproductive female body, symbolized by objects springing forth from bodies or hidden within them, as in the use of eggs, "false" consciousness, and other metal/technological additions to the body. Indeed, the presence of technology only seems to heighten anxiety over the permeable and eruptive qualities of the body. There is a deep fear of the cyborg, a fear of the grotesque possibilities of the mix of human and machine. Overall, the images evoke a deep fear and anxiety over human vulnerability in *any* environment.

While Aeon's femaleness suggests vulnerability, environmental instability, and chaos, Aeon, as a female, ultimately comes across as powerful. For example, it is impossible, in the world of *Aeon Flux*, to tell who is right or who is wrong. At the same time bodies are "female," the female is the contaminant, entering the body. The ambiguity of her character makes her highly "open" as a text—the viewers can place their own interpretation on her. Viewers might see her as a woman who never bows down to anyone, who sets the man in his place. She is an "unruly" woman, whose liminal status signals destruction.

At the same time, the concept of destruction has no meaning without the concept of existence. Ultimately, in a world of chaos, imminence is stressed. If nothing can be controlled and end results are unknown, uncontrollable, the best you can do is simply to exist. Existing as an action heroine, in motion, Aeon at least connects to the idea of the here and now, appealing to the visceral in all of us. Aeon, who like her name Flux, goes with the flow, seems wiser than Trevor, the rational man, who seeks to control what can't be controlled.

CHAPTER FIVE

Buffy the Vampire Slayer and the Body in Relation

Based on a 1992 film of the same name, *Buffy the Vampire Slayer* quickly became a cult hit after its 1997 premiere on the WB Network. The series, now airing on UPN, is in its seventh season, has spawned a spin-off series, *Angel*, and a wide range of collectibles such as t-shirts, novels, comics, and action figures. The numerous fan-run web sites and abundance of fan fiction on the web point to the series' intense popularity among its fans. In a *TV Guide* on-line poll of viewers, fans ranked *Buffy* as the number one show of all time.[1] The series' popularity extends to critics and academics. *TV Guide* critics included *Buffy* in its list of the 50 greatest TV shows of all time, and the series was nominated for best drama series in 2001 by the American Film Institute.[2] An academic on-line journal, slayage.com, publishes scholarly criticism focused on the show, while an entire academic conference on *Buffy* was hosted by the University of East Anglia in the United Kingdom in the fall of 2002.

Buffy's popularity among fans and critics alike points to its richness and complexity as a text. While there are many important elements to consider about *Buffy*, including the series' representation of female power and its status as a "feminist" text, this chapter is primarily focused on examining the series' concern with materiality. Like most heroic narratives, *Buffy* is ultimately about what it means to be alive in the world. Heroic narratives and the images of bodies within them mediate a particular ontology and thus function as blueprints or guides for how one ought to live and behave in the world. It is important to remember that, although most male-centered action texts focus on the body, the body is important primarily as a *sign* of heroic subjectivity: It is a means to an end, not the end itself. The individualism and images of bodily hardness found in most male-centered texts convey a sense that the hero is unique and separate from others. This suggests that the subject, the hero, is isolated and alone, as well as communicating a sense of the hero's superiority to others—a sense that, in turn, works to justify the hero's domination of others and his environment. Since we are alone, such texts imply that the only meaning we can take from the world is based on our concrete actions upon that world. Our actions define us, but the material realm is once again a means to the end of our self-expression.

However, *Buffy* frequently suggests the body is an end in itself, foregrounding both the pleasures and dangers of the material world and continually asking the question of what it *feels* like to be in the world. The series' narrative and textual strategies foreground bodily experience for both its characters and audience. *Buffy*'s textual strategies like metaphor and camp are ways the series works to make affect—bodily sensation and feeling—material and encourages viewers to see themselves as participating in a shared cultural imaginary. The series' narrative and moral universe likewise stresses the ways in which the experience of materiality, of being in the world, is always communal and relational. Its complex narrative is a coming-of-age tale focused on relationships and the realm of the body, particularly sexuality and desire. *Buffy*'s representation of sexuality and desire as dangerous expresses Buffy's often contradictory relationship to her own body, suggesting the ways in which the experience of the body changes in relation to the material world. *Buffy*'s morality and ethics are similarly based around an understanding that the body of the hero is not discreet but connected to and formed in relation to the bodies of others.

Camp, Metaphor, and Emotional Realism

Buffy's emphasis on affect and community is apparent in its complex narrative. In the original film, Buffy, a popular high school cheerleader, encounters a strange man who informs her that she is the new "Slayer." He explains that in every generation a new Slayer is born, gifted with extraordinary physical strength. The Slayer's sacred duty is to slay vampires, aided by a guide and mentor called "The Watcher." Buffy is disbelieving at first until the town and her school are taken over by vampires. The movie ends with Buffy's triumphant defeat of the vampires at the prom.

The series begins where the film left off. Buffy, expelled from her old high school in Los Angeles, has moved with her mother to a small town called Sunnydale. Though she is hoping to leave her past behind her, it turns out that Sunnydale is located directly over a "Hellmouth." The Hellmouth is a paranormal hotspot, "the undead's favorite party town," a center of mystical convergences, which means that weird occurrences are frequent. In Sunnydale, Buffy acquires a new Watcher, Giles, and a group of friends who help her in her Slayer duties.

Over six seasons, Buffy's cast of friends has expanded and evolved. Her helpers have included Willow, a computer geek who later becomes a powerful witch; Xander, an average guy with a heart of gold; Cordelia, a pampered princess; Oz, a werewolf; and Angel, a vampire with a soul. In the later seasons, other helpers appear, including Anya, a former "vengeance demon" trapped in the body of a high school girl; Riley, Buffy's college boyfriend; Tara, a witch who becomes Willow's lover; and Spike, a vampire who is "neutered" by a government-implanted electronic chip that makes him unable to harm humans. In the fifth season, Buffy also magically acquires a teenaged younger sister, Dawn.

The location of Sunnydale over the Hellmouth means that Buffy is rarely free of her role as a Slayer. Strange and unnatural occurrences happen all over town. Vampires regularly arise from the graveyard to terrorize the townsfolk, although, curiously, few residents seem to be aware of the numerous murders. However, Buffy's activities are not confined to killing "vamps," as she refers to them. Her opponents include demons, monsters, mummies, zombies, and witches, with each season featuring a metanarrative revolving around a different arch-nemesis.

Season One saw her defeat The Master, a powerful ancient vampire trapped beneath the town. In this battle, Buffy dies but is resuscitated by Xander. In Season Two, Buffy faces off against her one-time lover, Angel, and throws him into hell. In the third season, Buffy is pitted against another Slayer, Faith, who turns to the dark side and allies herself with Sunnydale's evil mayor, who plans to become a demon and take over the world during Buffy's high school graduation. She foils his plan and, in the process, blows up the school.

During Buffy's first year in college she battles a part-human, part-demon, part-machine villain created by a secret government organization called the Initiative. In the fifth season, in order to protect a mysterious energy source called The Key from an evil god named Glory, a sect of monks transform the Key into Buffy's sister, Dawn. Buffy's struggle to protect Dawn from Glory ends when she sacrifices her own life to save her.

The sixth season breaks many of the patterns established throughout the previous years. It focuses on the internal psychic demons of its protagonists rather than external monsters. Buffy is magically brought back to life by her friends but suffers from depression and anger at being ripped out of heaven to return to her hard existence. She is unable to emotionally connect to her friends and her sister and enters into a violent, destructive sexual relationship with Spike, her former enemy. Meanwhile, the family unit she has constructed throughout six seasons with her helpers disintegrates. Giles, her Watcher and father figure, leaves Sunnydale while the other characters begin to implode. Tara dies; Anya returns to being a vengeance demon, while sweet, goofy, and lovable Willow develops an addiction to magic and becomes the season's "Big Bad" going on a murderous rampage and attempting to destroy the world. Despite the character implosion, however, the season ends on a positive note with all of the characters overcoming their personal demons and experiencing a kind of "rebirth." Willow is stopped and saved by Xander's unconditional love for her; Spike, intent on destroying Buffy after she spurns him, is given a soul by a mysterious shaman; and Buffy is reconciled with Dawn and realizes she loves the world after all.

The characters and situations in Buffy resonate strongly with viewers, judging by on-line fan activities. *Buffy* inspires such an intense interest that fans seek to create more fantasies around *Buffy* than any other series. For example, at the time of this writing, "Fanfiction.net," the largest archive on the web for fan-written fiction, houses over twelve hundred stories written about characters and situation taken from the *Buffy* universe. The next closest series, *The X-Files*, has only around 3,500 stories in contrast. (*Angel*, *Buffy*'s sister series, is fourth with 2,300.) Although there are many possible reasons for *Buffy*'s strong fan appeal, one is certainly the way its textual components invite the viewer into an almost physical relationship to the text.

One factor that may inspire fan devotion is *Buffy*'s soap-like narrative structure, in which story lines and characters evolve over time. This structure allows viewers to become emotionally invested in the changes experienced by characters over time in a way that closed narratives do not. The extreme sincerity with which the actors portray the emotional world of their characters conveys a sense of the text as being extremely "realistic." As Buffy's mother tells her: "Everything is life and death to a teen-age girl." Similarly, in the early seasons, the teenaged cast talks about "relationships," love, and dating with an intensity that is, on the one hand, grandiose and absurd and, on the other, perfectly portrays the intensity of feeling experienced by real-life teens. As in *Nikita*, this intensity is enhanced by other textual components, such as the prominent use of a foregrounded soundtrack.

Although similar in structure to "realistic" shows like *ER* or *NYPD Blue*, which also feature an extended narrative, *Buffy* is clearly placed in a fantasy realm. Notwithstanding the otherworldly nature of most of the story lines and characters, the viewer is perpetually coaxed into experiencing intense emotions by the text. The plot might be crazy, but the emotions are real. Like a Grimm's fairy tale, the heightened sense of drama and horror expressed through fantastical situations rather than realistic reflection creates a particularly intense viewing experience. In her study of fans of *Dallas*, Ien Ang argued that *Dallas* viewers knew that the melodramatic occurrences of the text were far-fetched, but they still connected to the sense of emotion conveyed by the story. Ang observes that "the pleasures of *Dallas* consist in the recognition of ideas that fit in with the viewer's imaginative world. They can "lose" themselves in *Dallas* because the program symbolizes a structure of feeling which connects up with the ways in which they encounter life."[3] Thus, while the story lines and characters in *Buffy*, like those in *Dallas*, are farfetched, they still convey a sense of what Ang calls "emotional realism."

Buffy's consistent use of metaphors-made-literal works similarly. The heightened emotions of outlandish experience—falling in love with a vampire, for instance, or sleeping with your true love only to have him turn into "a monster"—model experiences with unrequited love with which most people can identify. They also express the common feeling that our dates are monsters, and we must be losers to have picked them. The episode "The Pack" similarly evokes the terror of being bullied through its depiction of group of unruly students who become possessed by the spirit of hyenas, eating the high school principal and the school mascot. The feelings of loneliness and isolation many teens experience are made concrete in "Invisible Girl" when a teen literally becomes invisible after years of being ignored. In "Life Serial" the feeling that the nine-to-five work day is "endless" is made literal when Buffy finds herself trapped in a repeating time loop at work. Such "monster metaphors" make the world of inner feelings concrete. They present psychic structures of feeling in a literal fashion, giving material form to ephemeral feelings.

In addition, many of *Buffy*'s "monster metaphors" are based on imagery that is drawn from commonsense cultural perspectives and cultural debate. For example, in the episode "Witches," Buffy tries out for the cheerleading team and befriends a cheer-obsessed teen named Amy, who is desperate to achieve the same high school successes her mother had. Amy seems to be suffering from extreme pressure put upon her by her mother. When Amy and Buffy are put on the alternate list for the team, mysterious accidents happen to the first string: one spontaneously combusts; another loses her eyesight; while still another's mouth is removed. Buffy and her friends immediately suspect Amy is the witch, but when they investigate her house, they discover that her fearful trembling mother is actually Amy. Amy's mother, so intent on living through her daughter, has literally stolen her daughter's body. "Witches" depicts the notion of mothers living through their daughters, an image that is present in current conversations over child beauty pageants as well as the Texas cheerleader murders. It also conveys a probably common feeling among many that "my mother is taking over my life."

Another episode draws from cultural discourse over the privileging of athletes in high school and a common understanding of Eastern Bloc countries' experimentation with steroids. Sunnydale swim team members rule the school, are given passing grades, and allowed to do whatever they want by the doting principal. One jock attempts to sexually assault Buffy and then claims that she provoked him by dressing provocatively. She is scolded by both the coach and the principal for her behavior. Ultimately, Buffy discovers the team is being given a mix of steroids their coach has been pumping into the steam room. The steroids aren't *just* steroids, however, but are mixed with a concentrate discovered by the coach after the fall of the Soviet Union, which had been experimenting on their swimmers with fish DNA. Unfortunately, the combination eventually turns the swimmers into man-eating creatures.

"Reptile Boy" relies upon an understanding of the commonsense fact that frat parties are places of danger for girls as the basis for the story of Buffy and Cordelia's adventures at a frat party. Teens who sneak out and go to frat parties never know what kind of horror or violence they might encounter. Seeking romance, girls might be used by the frat boys for their own vile purposes. The frat boys are indeed seeking something other than romance from Cordelia and Buffy. In a twist which stems from the existence of secret fraternity rituals, it turns out that these frat brothers worship a demon who lives in the basement: a big, snaky monster (the phallic reference, I am sure, is intended) who feeds off the flesh of girls. Buffy and Cordelia are offered as sacrifice. In exchange for making sacrifices, the frat boys, whose broader fraternity members include high-ranking government officials and millionaires, are guaranteed wealth and success.

The deep cultural concerns the series taps into became even more poignant after the nation was rocked by the high school massacre in Littleton, Colorado. It is telling that, in response to this tragedy, the WB withdrew the episode "Earshot" scheduled to run on April 27, 1999, in which Buffy thinks she overhears Sunnydale teens planning a massacre at the high school; the villain turns out to be the school cook. The WB later also postponed the airing of the third season finale, which featured the graduating class in an Armageddon-like battle with the monstrous town mayor. During the battle Principal Snyder is eaten, and the high school is completely destroyed.[4] Obviously the withdrawal itself reflects larger cultural debates about whether the media "causes" teen actions, a particularly intense debate following the events in Littleton and others in which teen rampages were believed to be modeled after films.

Buffy's imagery is thus both shaped by and extends upon an existing cultural imaginary in which viewers already participate. *Buffy*, as a text, is thus easily accessible and may be read as linked to our culturally informed consciousness. Viewers may be said to share an internal connection to these

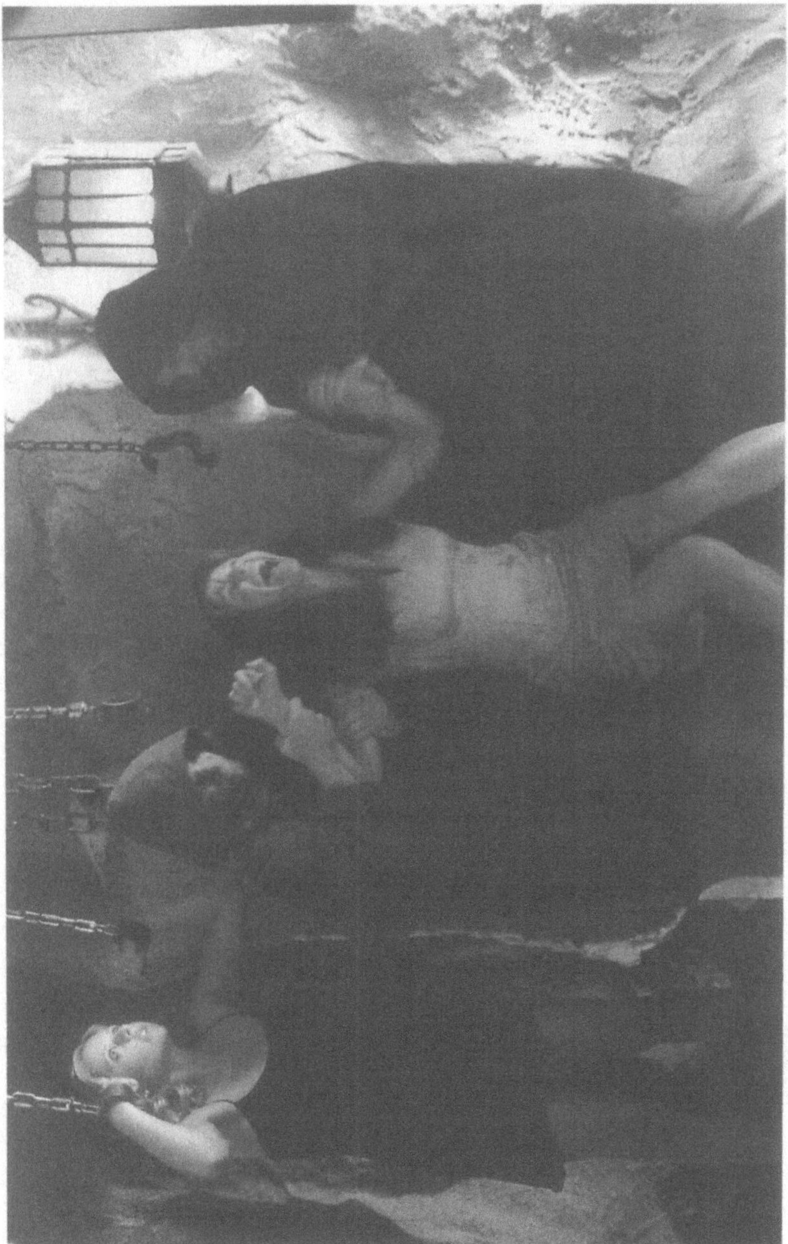

Figure 5. Buffy and Cordelia are attacked in "Reptile Boy."

metaphors. The series' use of camp and self-reflexivity works similarly to "ground" the viewer into the imaginary space of the text.

Camp is the conscious formation of a performance which displays that consciousness to the viewer. Camp is most evident in *Buffy*'s witty dialogue, which frequently references movies, people, urban legends, and folklore outside of the text in a self-reflexive way that makes the characters appear to be speaking from and sharing the "real world" of the audience. Camp thus works to undercut the audience's distance from the text by signaling that the producers "get" the joke: They understand the absurdity of the situation or the dialogue just as well as the audience, creating a sense of closeness, of a shared world between viewer and producer. (This assumed intimacy is reflected in the way fans constantly refer to the series' creator Joss Whedon as simply Joss.)

The dialogue, for example, often refers to common bodies of knowledge and forms of expression. For example, in "Reptile Boy," Buffy is saddened by the fact that Angel is never around except when there is danger. Feeling rejected, she goes out to a frat party. When Angel learns of this he is outraged. Willow scolds him for his neglect of Buffy, saying: "You're gonna live forever and you don't have time for a cup of coffee?" Similarly, when Angel and Buffy fight, Angel complains, "Whenever we fight, you *always* bring up the vampire thing." These exchanges indicate a commonsense awareness of how couples fight. As in many relationships, Angel's girlfriend repeatedly brings up the same slight. Moreover, the line equates Angel's vampirism to another guy's forgetting an anniversary. It also places being a deadly vampire on the same level as a thoughtless boyfriend—another joke. Camp thus deals with the absurdity of the plot by acknowledging the absurdity and then moving past it, encouraging the viewer to suspend disbelief and enter the world of the text.

Camp also creates a personal relationship to the text because of the way it implicates the viewer in the text. Dialogue like that in "Reptile Boy" or "When She Was Bad" turns a critical eye back on the viewer—who among us has not had a similar fight with a significant other, for example? Similarly, scenes such as those in which characters argue over who was the "better James Bond"—Sean Connery or Roger Moore—pull the viewer out of the story into the "real world," causing the viewer to witness her own ability to identify with (be mocked by) the plot, and then returning her back to the story. In such moments, not only do we watch the story, we are part of it.

Similarly, characters occasionally speak for fans. For example, the series' sixth season generated a lot of fan debate over its explicit presentation of sexuality. The writers acknowledge this in "Entropy" in which a trio of geeks witness Spike and Anya having sex on videotape and wonder, like the fans, if they are watching porn. James Marsters, who played Spike, received a great deal of fan notice due to the racy sex scenes of the season. The writers acknowledge this fan perspective through the character of Andrew, who watches the video and comments breathlessly, "Spike is so cool."

In any case, the camp references abound. There are references in the dialogue to *The Planet of the Apes*, *Amityville Horror*, *Witness*, and *Star Wars*, to name just a few. For example, in one episode the villainous vampire Spike encounters Angel and is outraged that he has become good. Spike, who was made into a vampire and previously mentored by Angel, rages: "You were my *sire*, man! You were my Yoda! You Uncle Tom!" Spike draws upon popular culture references to express their relationship of mentor and protégé and to critique Angel's sympathy for humans as a betrayal of his own kind. This line is funny in part because of imagining the monstrous Spike taking in a movie, but it is also humorous because it constructs vampires as being an oppressed people. The fact that they are so clearly *not* oppressed, simultaneously works to poke fun at "political correctness" as well as Spike. A similar line occurs when Angel tells Buffy he hates it when she calls him a vampire. She replies sarcastically, "What would you prefer? Undead American?"

Popular culture references are useful because they also work to turn the critical eye back at the audience. One particularly interesting reference is made in this first encounter with Spike when Angel tries to convince Spike that he is indeed evil. He says he has adopted a "puppy dog, I'm all tortured act" in order to fool Buffy. Spike laughs and says, "People still fall for that Anne Rice routine? What a world!" He is ridiculing Anne Rice's famous vampire series which features the vampire Louis as an ambiguous, highly romanticized creature. He is often tortured by the dichotomy between his need and desire to kill and his higher, spiritual nature. He sits around and ponders life, a kind of Hamlet of the Undead, who seems to never be able to make up his mind.

Spike's dismissal of Rice's notion of vampires deconstructs the idea of vampires as noble, suffering beings and instead positions them as truly bad. It mocks viewers who might hold a more romanticized opinion of vampires. The line also acknowledges Rice's influence in the construction of Angel, who, like Louis, is a suffering vampire. Finally, Spike's comment displays his self-awareness of his position within pop culture as well as demonstrating the same knowledge on the part of the series' producers.

Similarly, in one episode, Buffy is betrayed by an old friend who has discovered he is dying of cancer. He tries to turn her over to Spike in return for being made into a vampire. Meanwhile, he is aided in his efforts by a group of people who worship vampires. The vampires in which these people believe and refer to as "The Lonely Ones" are clearly patterned upon Rice's model of the lonely, glamorous, and ultra-humane vampire. The vampire groupies dress like vampires and congregate in a night club which is decorated with a vampire theme. Vampire movies play non-stop on the club's televisions.

The presence of television is not accidental. The groupies have obviously drawn their image of vampires from popular culture, following Rice and movies. In one scene, Buffy's friend mouths line after line of dialogue from vampire films. When he meets Spike, he tells him that he is not following the script correctly and begs him to say the expected lines correctly or "it won't be any fun." Spike is appalled.

Ultimately, this romanticized view of vampires is revealed as naive; the inhabitants of the club almost meet a grizzly death at the hands of the real vampires. However, the vampires themselves are not immune to popular views of vampires. Angel, entering the club, is disgusted by the naiveté of the clientele. He criticizes them for not knowing anything about how "real" vampires live or eat or dress. Angel is taken aback when a groupie passes him dressed in the exact same outfit as his own.

This scene puts the viewer on notice that *Buffy* is itself presenting an interpretation of "vampires" and is self-conscious of that fact. It also raises questions of vampire identity: What is "real," What is pop culturally induced? Even the vampires in the series may be influenced by the media's depiction of them. The referencing of popular notions of vampire behavior is interesting in that it works to define what vampires are "really like" in the "real world" of Sunnydale. Their true identity is Evil; at the same time they are just like people, enjoying media culture and even defining themselves in relation to popular images. The simulacrum has become more real than the real; how can the viewer scoff at the image on the screen as being "unrealistic?" Once again, camp works in a twisted way to place the "reality" of the series in the same space as that of the series' producers and viewers.

Far more than most shows, *Buffy* breaks down the separation between viewer and text. Camp, the use of metaphors made literal, and a narrative that is concerned with community, relationships, and feeling are rhetorical devices which urge the viewer to relate to the text in specific ways. Although the fantasy text does not show us what the world is "really" like, the viewer is consistently urged to recognize the text as highly "realistic" because *Buffy* offers an impression of what the world *feels* like. This feeling is both constructed in and extends upon an already existing cultural space of the imaginary. Participating in the text allows viewers to connect to this sense of a larger imaginary community.

While viewers of any text can be said to be participating in a shared imaginary community, this experience is perhaps more intense for *Buffy* fans because of its orientation towards concretizing feeling and the sensations of the lived body. Buffy's relationship to her body—its desires and feelings—expresses the series' concern with depicting what it feels like to *be* a body in the world. Her relationship with her body is also important to consider because of how it contributes to Buffy's construction as a hero. Carol Clover has noted that, of all forms of media, horror, and pornography are specifically designed to arouse bodily sensation in the viewer. Buffy's narrative actually blends qualities from these two genres in that its two most consistent themes are horror and sexuality. Among the many "horrors" encountered in growing up in the world of *Buffy*, sex is repeatedly depicted as one of most deadly. The eroticism of the text clearly makes the body central to the text and most probably works to inspire viewer feeling and textual engagement. It also frames Buffy's heroism and physical power in complex and often contradictory ways.

Justifying the Fight: Monstrous Men and Sexuality

Joss Whedon consciously created Buffy as an antidote to the typical girl victim of slasher films: "I started feeling bad for her. I thought, you know, it's time she had a chance to take back the night."[5] Similarly, a second-season full-page ad for *Buffy* features Buffy crawling over a table with her stake in hand. She is staring straight out at the reader: we are apparently her vampire prey. The copy says "Not exactly your average damsel in distress." The ad refers to the fairy-tale sensibility of the helpless damsel and mocks it, replacing it with an image of the "new woman," who, while sexy, threatens us with her stare and her stake and appears to be crawling off the page to hunt us down.[6]

Whedon's comment about "taking back the night" and extratextual framing like this ad are extremely interesting because they assume female sexual victimization as the norm. Likewise, although the series depicts a woman who is more physically powerful than anyone else, it is a power that is situated within a particular environment. Buffy's use of force is justified by the environment she occupies, in which "real" men are monsters while women are under consistent sexual attack and must constantly fight to keep themselves safe.

Like *Nikita*, the environment in Buffy echoes the terror of the gothic mansion of contemporary romance novels. The romantic theme is certainly framed by *Buffy's* extratextual artifacts like posters and t-shirts which feature images such as Buffy and Angel embracing in the center of a red heart, Buffy and Spike kissing, Willow and Oz kissing, and Buffy stepping through a heart pierced by a stake.[7] In *Buffy* relationships evolve around and through life and death circumstances, emphasizing their importance. Like a soap opera, relationships are central to both individual episodes and the metanarrative. A wide range of relationship issues, including mother-daughter dynamics and platonic relationships, is represented; however, *Buffy* places the greatest emphasis on love relationships.

Across all six seasons, love relationships are central to the characters' motivations and to the plot. For example, Xander loves Buffy, who in turn loves Angel, while Willow suffers unrequited love for Xander. The arc of the second season is built around Buffy's tortured relationship with Angel, a vampire cursed with a soul. When the two consummate their relationship, Angel loses his soul and turns evil. Spike betrays Angel to Buffy because after Angel turns evil, he steals the affections of Spike's girlfriend, Drusilla. Willow manages to restore Angel's soul but not before he has opened up a gate to hell. As a result, Buffy must kill him to save the world. Meanwhile, Willow has begun to date Oz, though their relationship is complicated by the fact that he is a werewolf. Xander dates Cordelia, but she dumps him after she and Oz discover Willow and Xander kissing. Angel is sent back from hell by The Powers That Be, but he risks losing his soul again if he and Buffy have sex. Eventually he leaves Sunnydale so she can have a normal relationship. The arc of the fourth season is built around the complications of Buffy's new relationship with Riley, an agent for a secret government agency dedicated to eradicating demons. Meanwhile, Oz dumps Willow when he realizes he cannot control his werewolf side. Willow is devastated until she meets Tara and comes out as a lesbian. Xander begins a relationship with Anya, a vengeance demon turned human. From the supernatural to the socially marginalized, all kinds of love have been explored.

In the fifth season, love relationships continue to be torturous. Buffy is abandoned by Riley and loses her mother but gains a sister, Dawn, for whom she must serve as a surrogate parent. Spike falls in love with Buffy, and his unrequited love for her becomes the central arc for his character, motivating him to do good deeds. Relationships take center stage during the sixth season. Xander and Anya plan on getting married until Xander leaves Anya at the altar. Tara leaves Willow because of Willow's increasing reliance on magic. Spike and Buffy begin an affair until eventually she admits to herself she is using him and dumps him. Anya and Spike seek solace in each other's arms but are caught on videotape by Xander and Buffy. Spike goes to apologize to Buffy but loses control and attempts to rape her. Tara and Willow reconcile, but Tara is shot and killed by a stray bullet soon after. A grief-stricken Willow attempts to destroy the world but is stopped by Xander's love for her.

While platonic love, like Xander's for Willow in season six, often plays an important and healing role, sexual relationships are almost always depicted as dangerous and deceptive and create an atmosphere of horror. Buffy's ill-fated romance with Angel, who turns evil after the first time they sleep together, exemplifies this rule. Likewise, in "Teacher's Pet" Xander has a crush on a teacher who is in reality a giant praying mantis that mates with and then eats virgin boys. In "Inca Mummy Girl," he falls in love with a mummy who sucks life out of boys with her kiss. In "I Robot, You Jane," Willow falls in love with a boy she meets over the Internet. Her lover turns out to be a demon trapped in the computer network who attempts to take over the world. It is important that there is never a point at which all characters are completely satisfied and balanced at the same time. There is always disequilibrium.

That this view on the dangers of sexual relationships is conscious on the part of the producers is made evident by the self-reflexivity of the characters. "I Robot, You Jane" closes with Buffy, Xander, and Willow discussing their failed attempts at love.

Willow:	The one boy that's really liked me and he turns out to be a demon robot. What does that say about me?
Buffy:	It doesn't say anything about you.
Willow:	I mean I thought I was really falling—
Buffy:	—hey, did you forget? The *one* boy I've had the hots for since I moved here turned out to be a vampire.
Xander:	Right. And the teacher I had a crush on: giant praying mantis!
Willow: (cheering up)	That's true.
Xander: (smiling)	Yeah, that's life on the Hellmouth!
Buffy:	Let's face it! None of us are ever going to have a happy, normal relationship!
Xander:	We're doomed!
Willow: (smiling)	Yeah!

Similarly, after failed attempts at dating the insect and the mummy, Xander broodingly tells Buffy that "I have the worst taste in women in the world. *Ever*." Buffy reassures him that the mummy really did care for him. He replies: "Yeah. But I think that whole sucking-the-life-out-of-people thing would have put a strain on the relationship."

Sex between men and women almost invariably leads to a transformation, betrayal, or revelation, and ultimately violence. For example, when Buffy sleeps with Angel, he becomes evil. Xander is almost strangled by Faith. Oz's sexual jealousy over Willow's relationship with Tara makes his werewolf side uncontrollable. In the fourth season, Buffy and Riley's love-making releases the repressed sexuality of a poltergeist, which imprisons them and forces them to continue to make love—literally fucking themselves to death. While lesbian lovers Willow and Tara had a healthy, loving relationship for years, there were ominous links drawn between lesbianism and the corrupting power of witchcraft. Tara's death immediately after reunion sex with Willow recalled a homophobic cliche and caused an outcry among many fans.[8]

Vampires have historically been used to signal sexual danger and, in a show about a vampire Slayer, clearly create a sexualized, dangerous space.[9] The vampire's bite is treated as a sex act: Riley's allowing himself to be bitten is treated as a sexual indiscretion; Buffy appears orgasmic when she is bitten by Angel. In "The Initiative," the scene where Spike learns he can no longer feed is consciously and comically played as a scene of sexual impotence. Spike enters Willow's dorm room, throws her across the bed, and turns up the music on her stereo to drown out the sound of her screams. The moment is highly suggestive of a rape as he pins her beneath him and tries to bite. He is unable to perform. In the next scene he sits hunched at the edge of the bed, and the dialogue shows how both characters react to the sexual nature of the unfulfilled bite:

Spike:	I don't understand. This sort of thing has never happened to me before.
Willow:	Maybe you were nervous.
Spike:	I felt all right when we started. Let's try again—Dammit!
Willow:	Maybe you're trying too hard. Doesn't this happen to every vampire?
Spike:	Not to me it doesn't!
Willow:	It's me, isn't it? I'm not the kind of girl vamps like to sink their teeth into. It's always like, "Oh you're like a sister to me" or "You're such a good friend."

Willow's relief at not being bitten is rapidly transformed into pique at not being found desirable. The vampires' ability to kill and bite is linked to sexual potency. After losing his bite, Spike is described as "neutered," "impotent," and even "flaccid." When he discovers he can hit Buffy, he claims he has his "rocks" back.

Certainly, danger is part of love for both male and female characters. However, this danger is most pronounced for the women on the show, most particularly Buffy, who is drawn to dating vampires. One episode even has Buffy dressed up as Little Red Riding Hood packing weapons in her basket. (Spike, her current love interest, calls himself the "Big Bad"), a self-reflexive joke on the part of the producers regarding the way Buffy is continually framed in an environment that is sexualized and dangerous. In the world of

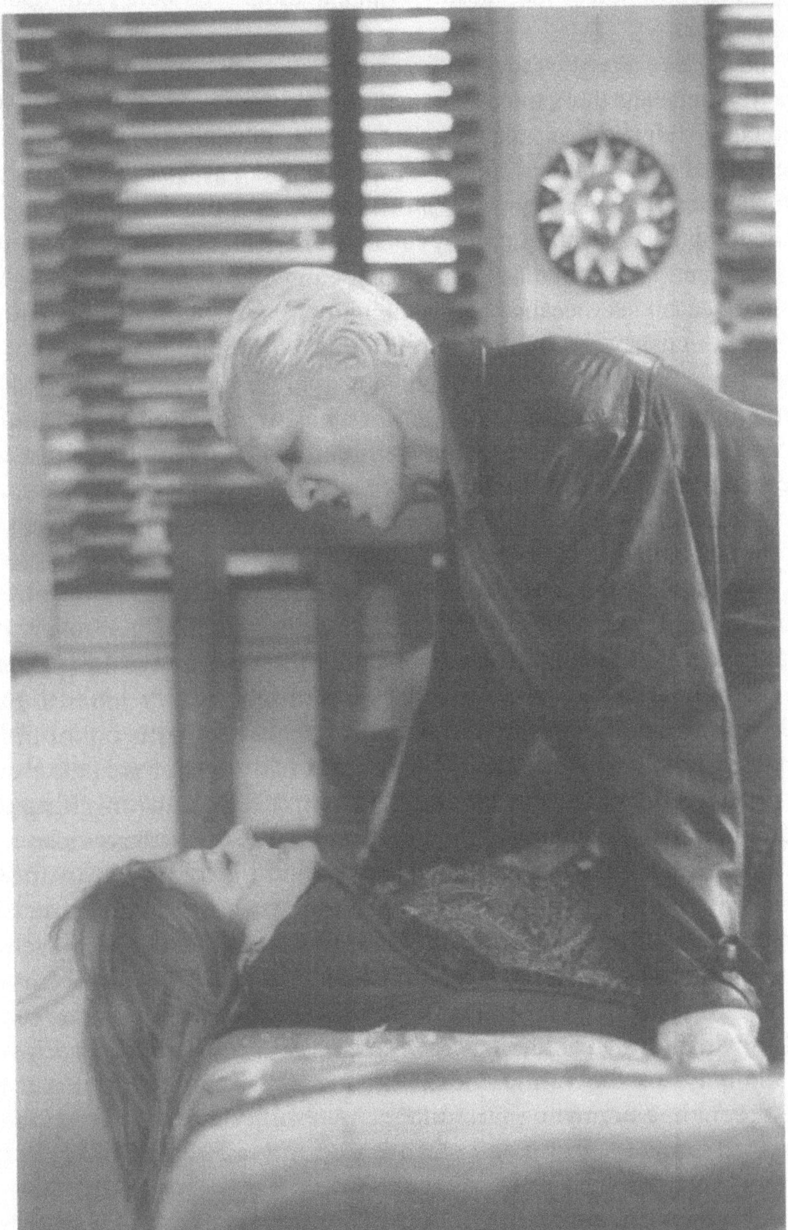

Figure 6. Spike attacks Willow.

Buffy, sexually potent masculinity is threatening.[10] To be a "real man" is to be a monster.

The series strongly suggests that most men are intrinsically dangerous and wild, and that this monstrous quality is desirable. Men in *Buffy* are unpredictable, depicted as dangerous at heart, and almost never what they seem. Even Buffy's father figure, Giles, has a dark past as "The Ripper." The one character who seems to offer a different view of masculinity is Xander. Although he is constantly upset by his awareness that he is not a "real man," he is frequently presented as heroic, possessing a heart of gold, and he is the savior of season six, because of his "feminine" capacity for caring. However, Xander is constantly finding himself in relationships in which an angry, powerful, or dangerous female dominates him. He also demonstrates homoerotic longing at times, for example, when he describes Spike as "strong and mysterious and sort of compact but well-muscled." It is important that as the one character who consistently refutes the rule of monstrous male, he is never presented as a sexually viable partner for Buffy, and his sexual choices reflect his impotency. He is not a "real man."

In addition to the presence of dangerous men, the environment itself is extremely uncertain. Buffy is continually confronting the question about the truth of identity. Much of the action involves the fact that evil has penetrated the realm of the everyday in an unrecognized form. Buffy's job is to restore order to the land, but to do this, she must recognize what is out of place in whatever form the evil takes. Evil may come in disguise, even in the shape of her friends. Buffy fights in a world where people are constantly changing. In addition to implying vulnerability, the constant flux of characters also means that even the familiar can become unknown and threatening. It is often unclear who are the villains and who are the protagonists. For example, the vampires have the capability to appear as humans. Although as Slayer, Buffy is supposed to be able to sense their presence, for some reason she lacks this ability. This means that she often finds herself relying on her fashion sense to judge by their taste in clothing whether they have been buried for ten years.

Similarly, of the various demons, monsters, and witches that appear, most seem to be normal on the surface. A substitute science teacher is actually a giant praying mantis; an exchange student is a 500-year-old mummy; a nervous cheerleader is actually a witch, while a high-school talent show contestant is an organ-eating demon. Buffy's college psychology teacher heads a nefarious project to create monsters for the military. Ben, a hospital intern with whom Buffy flirts, is the alter ego of the evil goddess Glory. Sixth season villains, Warren, Andrew, and Jonathon are D&D-playing nerds who reveal the banality of evil. Seemingly harmless, they attempt rape and commit murder.

In later seasons this instability of character occurs not just because of magic but is constructed as part of human nature. Oz learns he cannot control the werewolf inside of him; Buffy's boyfriend Riley, a "farmboy" type from Iowa, seemingly trustworthy and devoted to Buffy, cheats on her with vampire prostitutes—he pays them to suck his blood. Giles kills Ben. Dawn shoplifts repeatedly. Willow's developing witch powers lead her to take increasing risks with her often dangerous magical spells. She exhibits all the characteristics of an addict, even casting mind-control spells upon her friends. Mad with grief over the death of her girlfriend, she tortures and kills a man and tries to destroy the world.

It is vital that in this framework, nobody is impervious to harm. Everybody may be penetrated and affected at the core of their beings by external forces both magical and "real." A remarkable number of featured recurring characters die. Miss Calendar, a regular for the first two seasons, and Tara, a regular for three seasons, are both brutally murdered. Even Buffy's mother, a regular for five years, dies from a sudden brain aneurysm. The nonmystical nature of her demise places her death firmly in the realm of reality. In the fifth-season episode "The Body," the cast has to come to grips with their helplessness in the face of death: It is not something that magic or slaying can fix. The death of such stable characters contains the threat that even the five regular cast members may die.

The physical environment is likewise unstable. There are no safe spaces, and environment is often used to reflect the vulnerability of characters. Buffy's house, for example, is frequently used as a metaphor for her own body. Initially, her home was a safe space. The "normalcy" of suburban living is juxtaposed against Buffy's outside life that intrudes upon her desire for a more peaceful existence as well as her mother's demands for her to settle down. Her mother, unaware of her true identity, still treats her like a little girl. But the outside world of the streets and adventure is Buffy's true life, one which intrudes upon her mental and physical security in her house and also affects her ability to *be* the little girl her mother wants her to be. As the series progresses, her home is broken into more and more often; first by vampires then by monsters, demons, homicidal robots, and zombies. With this influx of evil, Buffy's mother is forced to recognize her true identity as Slayer, which also serves as her acknowledgment that her daughter is growing up.

The metaphor of the house also reflects the environment of sexual danger. This connection is reaffirmed in several pivotal moments between Buffy and her two vampire lovers. Vampires can only enter a house once they are invited. Angelus (Angel turned evil) is able to enter Buffy's house, and she must cast a spell to ward him off. The sexual nature of the metaphor is made even more prominent in later seasons, in which Buffy's willingness to invite Spike into her house charts the progress of their relationship. In "Shadow," Riley's sexual jealousy is spurred on by the knowledge that Buffy has never disinvited Spike from her house. In "Crush," Spike kidnaps Buffy and declares his love to her. Buffy, while clearly not scared by him physically, does appear to be emotionally threatened by his revelation and reacts to the sexual threat he poses by banning him from her house. In the final fifth-season episode, "The Gift," Buffy invites him in once more. Spike slowly, happily steps inside and says "Hmm. Presto. No barrier." The moment is a pivotal one in their relationship, as it is the first sign that Buffy will "let him in" to her life. "Smashed," in which Buffy and Spike consummate their relationship, juxtaposes images of a building smashing (complete with cracking walls and falling timber) around them as they make love, simultaneously suggesting the destructive nature of the relationship and the crumbling of barriers between the two.

The physical space of the series is thus uncertain, unstable, dangerous, and sexualized. In this way, *Buffy*, like *Nikita*, describes a similar journey to those of heroines of contemporary gothic romances in which the heroine must navigate dangerous spaces while dealing with the question of the "truth" of identity of those around her, particularly ambiguous and dangerous men.[11] Unlike the heroine of traditional romances, Buffy is a physically powerful figure, but her power is necessary because she is physically endangered, existing in a dangerous environment pervaded by threatening male sexuality. Buffy's relationship to her own body is thus contradictory.

An example of this contradiction can be seen in the second-season episode, "Halloween." This episode is interesting for the way it mocks adherence to strict gender roles while simultaneously reaffirming the notion of the world as dangerous for girls. In this episode, the students of Sunnydale are excited about dressing up for Halloween. Buffy tells Willow that Halloween is wonderful because it is the time to have a party where you "come as you're not." Thumbing through one of Angel's diaries, they discover a drawing of a beautiful woman garbed in clothes from the late eighteenth century. Buffy tells Willow about her desire to be a delicate, fragile lady with servants and beautiful clothes. Her desire is at least partly prompted by her belief that Angel would prefer a lady of his own times. Willow reminds her that there are nice things about existing in the present, "Like being able to vote."

At the same time that Buffy yearns to be "feminine," Xander is also suffering a gender crisis. When a classmate accuses Buffy of loose sexual morals, Xander tries to fight him. He needs to be rescued by Buffy, who quite causally twists the bully's arm, slams him against a drink machine, and gets a free soda. "Diet!" she exclaims in delight, underscoring her "femininity" and the fact that she, a tiny girl, has bested a bully and protected her male friend. Xander feels emasculated and is enraged at her.

The plot thickens when Buffy finds an eighteenth-century dress in a costume shop and decides to dress up for Angel. Unbeknownst to the students at Sunnydale, the owner of the costume shop is a warlock. He casts a spell which turns everybody wearing clothing from his shop into the character they portray. Buffy, dressed as a lady from the eighteenth century, becomes that lady. Xander, dressed as a soldier, is changed into a man of action. Once transformed, they act out gender stereotypes. Buffy is an exaggerated vision of femininity, fainting, screaming, and crying. She utters such phrases as "surely the men will protect us."

Her dialogue and actions are designed to contrast not only with Buffy's "true" identity as an action heroine but also to highlight the difference between old-fashioned notions of proper feminine behavior and more contemporary views. Ultimately, although recognizing a yearning that many girls probably have in regards to the "romance" of the olden times, "Halloween" shows the modern perspective as better. Buffy, in her old-fashioned state, is a helpless, boring drag who endangers everybody. She is nearly killed by Spike, who seems torn between eating and ravishing her.

"Halloween" shows characters adopting various identities but ultimately affirms that there is a "true" identity—at least for the female characters. For example, in an attempt to impress Xander and break out of her traditional role as the nice-but-plain-girl-next-door, Willow dresses up in a shocking, sexy outfit with black leather miniskirt, belly-shirt, choker, thigh-high boots, and make-up. However, she can not bring herself to appear in public dressed that way. At the last minute she throws a sheet over herself, going as a ghost, and turns into a ghost for the remainder of the evening. At the end of the episode, she removes her white sheet and tosses it aside, as if to say "enough of costumes." Tellingly, what remains underneath is her sexy girl dress suggesting that, although she has attempted to cover it up, she is, in truth, a sexy, sexual girl. Importantly, as she walks through the night, Oz sees her from his car and is love struck. Her true identity, once revealed, wins her the love and admiration of a man.

It is important that Buffy learns that being herself means that she is tough. She discovers that when she is not around to fight the bad guys, havoc ensues. Her powers are necessary. In this way, the episode reaffirms the progresses of the feminist movement, with its focus on providing all individuals room for self-expression and being themselves. At the same time, the "progress" made by feminism (a term which is never mentioned), is defined against an extreme stereotype of rich, white female passivity and of women before the movement as being pampered, rich, silly, and weak rather than the realities of the laboring, poor, non-white, oppressed people. Thus the "improvements" brought by feminism are presented as enacting the growth and empowerment of the individual rather than widespread social change, reaffirming feminist progress in an environment in which the ideology of domination is still in play.

The episode reveals that one reason traditional femininity is outmoded and that girls need to be tough and independent is that the world is a dangerous place. You need to be "tough" to survive. If Buffy finds herself having a horrible time and being useless, Xander is able to satisfy and reaffirm his masculine identity through acting out his fantasy of being a military man. Fantasy desires can be both bad and good, depending on the context. The environment in which the characters find themselves in *Buffy* is a threatening place in which it is better to be tough.

Thus, the justification for Buffy's strength is based upon the idea that the world is dangerous and tough and requires fighting. Individual choice to be what you are is embraced against a context which values the ability to fight. Ultimately, Xander gets to revisualize himself as ultra masculine, which highlights the fact that in "real life," he is *not* the masculine "ideal." However, his visualization of himself as tough, unlike Buffy's dreams of feminine weakness, does not backfire against him. He is reaffirmed, not chastened, for his inner dreams. The masculinist paradigm of competition is still very much in sway.

While critical of traditional femininity, the series suggests that sexual appeal remains a key component of female identity. The push to be oneself is important because it wins the love and admiration of men. Willow's "true" identity is shown to be a sexual identity and is seen through the admiring eyes of Oz. When Buffy confesses to Angel that she dressed up for him, he tells her that eighteenth-century ladies were boring, only thinking of their looks. He prefers modern women (hinting at a double standard by which women are supposed to be effortlessly beautiful while thinking about loftier things). Male approval is presented as central to identity for these women. While Willow comes out as a lesbian in the fourth season and has a long-term relationship with Tara, male approval continues to be central for Buffy, as almost every season she is placed in a love relationship which is integral to the season's plot and which works to define and shape her character.

In addition, it is interesting that female individuality is embraced for one very traditional reason: its positive effects on social stability. For example, "Beauty and the Beasts" draws upon the theme of domestic violence. Buffy, Willow, and another high school girl, Debbie, are each involved with a man with an "animal," violent side. One morning, a shredded body of a teen is found in the woods. The two immediate suspects are Angel, who has returned from hell but is violent and insane, and Oz, the werewolf. Oz, now in boy form, is heartbroken. Locking himself up the following night, he warns Willow angrily to get away from his cage. He recognizes his wild side and treats her nastily in order to protect her. Despite Buffy and Willow's suspicions about their own boyfriends, Debbie's boyfriend, Pete, a human boy, is actually the murderer. Concerned that Debbie was leaving him because he was not manly enough, he created a potion to enhance his masculinity. Like Jeckyll and Hyde, he is transformed from a normal, nice guy to a possessive, raging demon, who kills those who make him jealous, and beats his girlfriend for leading them on. After beating her, he turns back into a normal boy and cries into her lap, begging her to still love him. Eventually, Pete is able to transform himself without the drug. Cordelia asks in horror, "It was a *real* killing? He wasn't under the influence of anything?" "Just himself," Buffy replies darkly.

The allusion to battered-wife syndrome is obvious. Debbie has internalized her own oppression. She blames herself and thinks it is her duty to love and protect Pete. She tells Buffy that she is all he has and "he doesn't mean it"; "It's all my fault"; " I make him mad." When Buffy and Willow confront Debbie, Buffy tells her that if she wants to continue being a victim she can, but she owes it to everyone else to stop her boyfriend from hurting others. Debbie's protection of Pete backfires as he becomes even more violent and is ultimately killed.

Despite the fact that Willow argues against the belief "that every guy is in it for the chase," the episode explicitly presents men as animals or monsters at heart. While this "fact" is certainly debatable, it is perhaps more interesting to observe how the episode treats the matter of how to react to male violence. Significantly, Buffy and Willow are shown as having the right approach to men. They each chain up or imprison their boyfriends in order to protect the public good. Buffy chains up Angel; Willow shoots Oz with a dart gun. Debbie, who allows herself to be victimized, does nobody any favors. Her passivity and self-denial actually allow Pete's violence to increase and indirectly causes the death of several characters. If anything, this episode promotes a notion that girls must stand up for themselves, while "good boys" like Oz will voluntarily enter their cages when their time of violence arises. The contrast between male violence and female violence reiterates stereotypical thoughts about gender and aggression. Male violence is unthinking and animalistic, while female violence is protective of the self and of others. Female violence is purposeful violence, rather than just an expression of strong emotions or a need for individual power.

It is interesting to observe the role of women in the creation of *Buffy*. Series creator Joss Whedon's control over the series is well known. He controls the overall shape of the narrative, serves as executive producer and writes and directs many episodes. However, *Buffy* has a larger than usual number of women participating in the upper level of production, including co-executive producer Marti Noxon. Jane Espenson, Rebecca Rand Kirshner, and Noxon write and produce, while Noxon also directs. Female writers have accounted for a relatively large percentage (at least 44 of 122) of the episodes to date. It is not surprising that many of the female-written episodes, like Noxon's "Beauty and the Beasts," "Wrecked," "New Moon Rising," and "Wild at Heart" (as indeed, the entire sixth season for which she served as co-executive producer) are among those that most explicitly focus on the dangers of male-female relations and the relationship between power and desire.

It is important to note that the dangerous sexualized environment that Buffy creates is not unrealistic: currently women live in a world in which their physical safety is at stake every day and in which they are most likely to be assaulted and killed by their sexual partners. Like camp and literalized monster metaphors, the dangerous environment depicted on the show works to create a realism of affect. For female viewers it is easy to acknowledge the dangerous sexualized environment as one that feels "real"; for many women, it is real. However, the sexually dangerous environment functions on other levels as well, in particular, for the way it links sexuality to female power.

The Darkness of Power and Desire

In a recent interview, Joss Whedon said that if vampires couldn't get erections, *Buffy* would have lasted only 12 episodes.[12] Vampire potency clearly plays a large role in helping to create the dangerous sexualized environment in which Buffy finds herself. However, sexuality and its textual connections/equation to power and desire are central to the text in other ways. Buffy's character has been developed and defined through a series of sexual relationships, which are not only central to her personal storyline but usually play a large part in each season's overarching storyline. There is consequently a great deal of textual focus on what Buffy does with her body—her sexuality impels the story lines as much as her fighting ability, literally working to shape the universe of the text. Buffy's sexuality causes men to lose their souls, to seek souls, and almost destroys the world. Like many aspects of the text, the representation of Buffy's sexuality is contradictory. On one hand, *Buffy* expresses the reality of female desire and equates it to her superpowers, yet it also, for the first six seasons at least, has depicted her sexual desires as destructive and linked to darkness.

The series quickly establishes the dangerous and destructive nature of female desire when Buffy sleeps with Angel and causes him to lose his soul.

Although he is evil, she can not bring herself to kill him. The consequences are dire: Angel kills Buffy's teacher, Miss Calendar, tortures Giles, and almost destroys the world. Eventually, Buffy harnesses her physical and emotional desire for Angel when she makes the choice to kill him (whose soul has since been returned) in order to save the world.

After her affair with Angel, Buffy begins a process of self-repression, exemplified by her relationship with Riley. Saying that she is "so over" the "whole bad boy thing," she chooses Riley, a seemingly average guy from Iowa, who is actually the commander of a secret military operation. "Into the Woods," displays that even normal Riley has the propensity for dark actions; the relationship ends when Buffy discovers he has been paying vampire whores to suck his blood. Despite this fetish, however, Riley is constructed as not being dark *enough* for Buffy. While the show implies that one reason Riley and Buffy did not work out was that Buffy has repressed too much of herself as a result of her experiences with Angel, Buffy also remains distanced from Riley, because, as Spike says, Buffy needs a "little monster in her man." He tells Riley that no matter how low he goes, it will never be enough for her. Buffy's desire for monstrous men seems confirmed in season six when she begins an affair with Spike.

Buffy's affair with Spike is particularly interesting to examine because it articulates the series' contradictory stance towards female sexuality. Does Buffy require a monster in her man because her desire is so great only a supernatural being can equal it or because her desire is perverse and corrupt? The storyline is also interesting due to the extreme ways in which fans reacted to it and because of how it makes the series' link between sex and danger extremely explicit. Coinciding with the series' move from the WB to UPN network, the consummation of their relationship was shocking for both its graphic depiction of violence as foreplay and the history of Spike's character as a sexual predator. The relationship demonstrates the often contradictory way the text presents Buffy as sexual victim while simultaneously showing her as sexually aggressive and desiring. However, at least by the ideological closure available at the end of the sixth season, her desire continues to be represented as wild and overwhelming, linked to danger and darkness.

Spike stalks Buffy for much of the fifth season, sneaking into her room to smell her sweaters and steal her panties, raiding her basement for photos. He dresses his vampire girlfriend up like her. He has an altar in his crypt dedicated to her and a mannequin with a blond wig that he alternately fondles and uses as a punching bag. He commissions a robot built in Buffy's

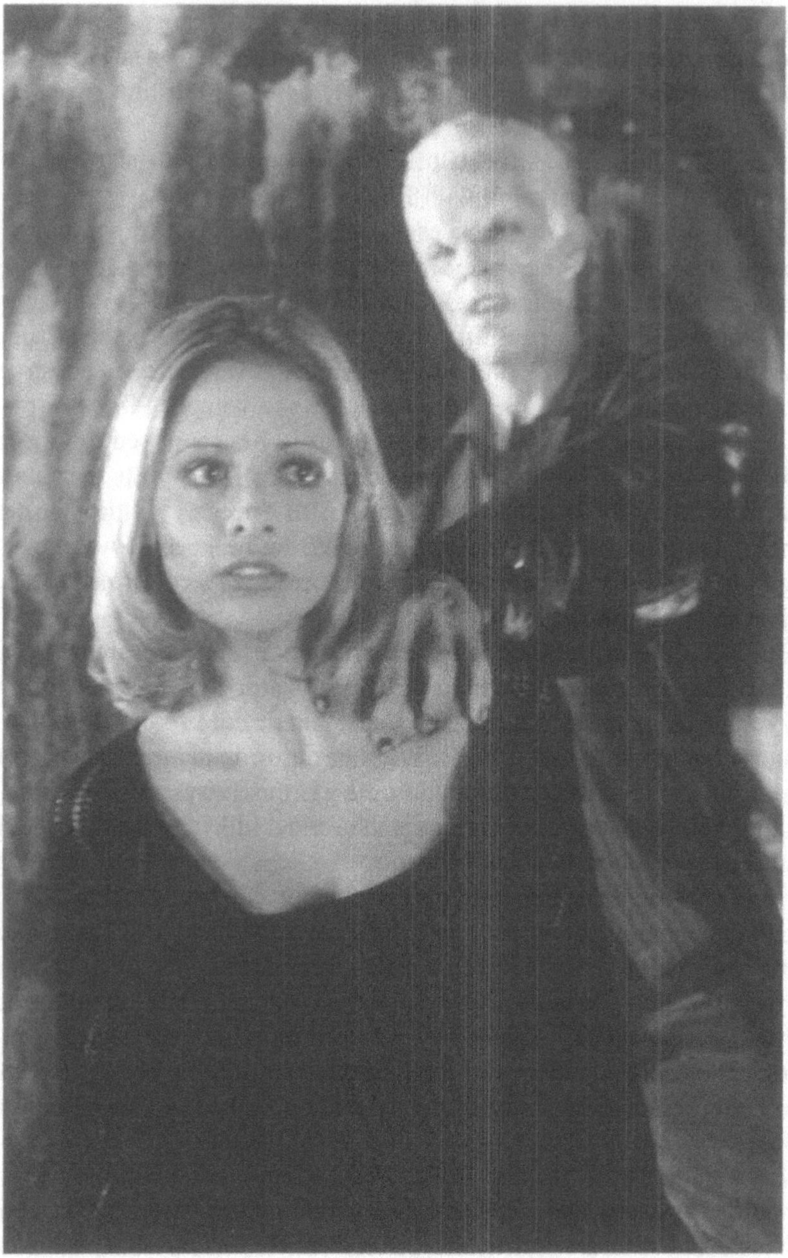

Figure 7. Buffy and Spike.

image that he uses as a sex toy. He even goes over to her house one night to shoot her.

The activities Spike engages in are horrifying, especially when you consider the results in the real world of such obsessive behavior. However, one of the problematic aspects of the series is that many of Spike's stalking activities are presented in a comic light—a questionable treatment of such a serious cultural issue that seems more in keeping with a *Penthouse* comic than a "feminist" television text. Moreover, we are meant to sympathize with Spike in many ways. For example, in "Crush," the fifth season's Valentine's Day episode, Spike kidnaps Buffy, chains her up, and tells her he will let his evil ex-girlfriend Drusilla eat her unless she admits she has some spark of feeling for him. Spike's actions are interpreted sympathetically by the writers of *Spectrum* magazine:

> Spike is clearly at wits end trying to impress Buffy, yet without a clue how to do it. Her vociferous denials regarding feelings for him aren't completely convincing....After numerous episodes of *Buffy* from the girl's point of view—men don't understand their needs, use them, act like jerks, et cetera—"Crush" is the other side of the coin, presenting Spike's frustration. The women don't understand him.[13]

One would think Spike's history as a killer and obsessed stalker would make his character irredeemable. However, Spike is increasingly presented in a sympathetic light. Certainly his love for Buffy is presented as more sincere as he devotes himself to helping her save her sister Dawn from the hands of the evil Glory. In "Intervention," Buffy is disgusted to learn that Spike has built a robot in her image which he uses as a sexual plaything but forgives him with a kiss at the end of the episode when she learns he allowed himself to be tortured by Glory in order to protect the secret of Dawn's identity. This episode, with Spike's later confession that he knows Buffy will never love him and his grief at the sight of her dead body in "The Gift," show his love to be less selfish than earlier presumed.

While Spike's gentler side was showcased early in season six, it is telling that Buffy and Spike consummate their relationship only after Spike learns that his chip no longer functions around Buffy and he is physically able to attack her. The two come to blows in an abandoned building in "Smashed." The fight clearly serves as foreplay: With the building crumbling around them, Buffy abruptly ends the fight with a kiss before shoving Spike and mounting him against the wall. The house collapses and the two plummet to the basement floor. Later encounters involve the destruction of Spike's crypt, sex in various public places, and bondage games.

Their desire for each other is linked to physical pain. Trying to figure out Spike's attraction for her in "Crush," Buffy says she beats Spike up a lot and "That's like third base for Spike," and she later tells him in "Smashed" that "you're in love with pain." But the love of pain is not one-sided: Spike replies, "Hello. Vampire. I'm supposed to be treading on the dark side. What's your excuse?" Buffy hits Spike; Spike hits Buffy. Both clearly enjoy the s/m aspect of the relationship, but while Spike apparently likes being beaten up, it is unclear if Buffy wants to dominate or be dominated.

Physically, Buffy is able to dominate Spike, but their romantic scenes are nonetheless laden with traditional gender ideas. Buffy is the "good girl," denying her true desires. "Smashed," "Gone," "Doublemeat Palace," "Dead Things," and "As You Were" all depict Buffy's attempt to fight her attraction to Spike. She continually tells him she will never do it again, but, invariably, she returns to him. Her sexuality is framed as explosive; after kicking him out of her house in "Gone," she later attacks him in his crypt, flinging him against the wall and ripping off his shirt, a wild animal. Spike, able to speak his desire, is the verbal aggressor in the relationship as a whole. Buffy's desire remains unspoken until it erupts uncontrollably with negative consequences.

Similarly, the morning-after scene in "Wrecked," resembles a wrestling match in which we learn that no really means yes. When Buffy tells him they are through, Spike, outraged, grabs her and sticks his hand up her skirt. Buffy enjoys it for a second, then pushes away from him

Buffy: (struggling)	Stop!
Spike:	Make me!
Buffy:	No!
(they kiss)	

Spike reaffirms the danger/sex connection, telling her: "I knew it! I knew the only thing better than killing a Slayer would be f—" before he is cut off by an outraged Buffy. The text also is rather obvious in drawing a parallel between Buffy and Spike's relationship with a dangerous drug, as it mirrors Willow's increasing reliance on dark magic. Spike even tells Buffy "I'm in your system now. You're going to crave me like I crave blood. And the next time you come calling, if you don't stop being such a bitch, maybe I will *bite* you."

The show is certainly self-aware about the dangerous line it treads. "Seeing Red," in fact, offers commentary on the patterns of violence and lack of communication occurring in Buffy and Spike's relationship. In this episode, Buffy has cut off her relationship with Spike. In an attempt to get her to admit what he thinks are her true feelings for him, he ends up sexually assaulting her.

Although self-conscious about the ways in which such patterns of interactions can culminate in violence, the text nonetheless is problematic for the way it handles the issue of rape. The attack is clearly supposed to be a Very

Bad Thing. It is filmed in a truly horrifying way—Spike looks like a huge black spider crawling up Buffy's back, and she is begging him in tears to leave her alone. However, it is disturbing primarily because it is easy to see the assault from both sides. It is quite clear that his actions are bad—Spike himself recognizes the depth of his transgression and is guilt-ridden. But Spike's point of view is poignantly apparent: He has been used by Buffy for sex, and she appears to want to erase the reality of what has transpired between them. He is suicidal and seems to literally almost suffer from a psychotic break because he is so desperate for Buffy's love. Moreover, given the context of their relationship all season, in which Buffy says no but means yes and in which fighting is foreplay, it is easy to believe, as Spike does, that she is simply denying her feelings.

This sense is furthered by how quickly Buffy seems to forgive Spike. In the very next episode she takes her sister Dawn to him for protection and is visibly upset when she learns he has left town: Apparently Buffy is not too upset by the attempted rape. While the scene has many artistic merits, for example, spurring Spike's realization that he is trapped between being a monster and a man, politically it walks a very thin line in dealing with a real social problem in that it urges the viewer to sympathize with the attacker's point of view.

The often uncomfortable way the series ties sexuality to violence leads to a reading that the series is merely replicating patriarchal ideas about female sexual passivity. "Ace," a reviewer from www.mightybigtv.com (now Television Without Pity), saw "Smashed" and "Wrecked" as failing in *Buffy*'s feminist message:

> I personally don't find the annoying, physically forceful cretin of the bodice-ripper romance to be a romantic ideal. I can't believe that what used to be my favorite show is forcing me to sit through a scene implying that our heroine is a little lady who just needs to be held down and forced into "letting down her barriers."[14]

Certainly, the transformation of Spike from villain to comic relief to romantic lead suggests that stalking and obsessed love are acceptable and even rather romantic. While the series' producers and the text itself frequently indicate that the Buffy/ Spike relationship is unhealthy, many fans are certainly reading it as a romance, judging by the vast number of "Spuffy" sites that emerged on-line. In a recent *TV Guide* on-line poll Buffy and Spike won "favorite TV couple," with 35% of the votes, ahead of *Friends'* Chandler and Monica (19%).[15] Spike also won 83% of the votes in Zap2it's ranking of Buffy's boyfriends.[16] Finally, www.jamesmarsters.com, a fan-run web site dedicated to the actor who plays Spike, crashed after the airing of "Wrecked" because of too much traffic.

The popularity of the relationship among fans demonstrates how sexualized violence so clearly ties into our cultural mythologies about love. However, while it is easy to interpret the text and fans as simply buying into myths of female victimization, the text is more complex. The sexual relationship between Spike and Buffy is undoubtedly problematic, but it also offers commentaries on the ways that power and desire are intertwined.

For example, Stephanie Zacharek interprets the violent but consensual sex depicted in the episodes "Smashed" and "Wrecked" as positively representing female desire. Following Camille Paglia's notion of sex as will-to-power, she applauds the series for detailing Buffy's desire. She also makes a valuable point about the very *messiness* of the sex depicted in "Smashed" and "Wrecked" as positive.

> The past two episodes of the show have been more sexually frank without being particularly explicit, and downright more exhilarating, than anything I've seen on TV in recent years....I can't think of a recent movie that dealt with the emotional risks and dangers of sex in such a startling and affecting way...they've so straightforwardly defined the usually blurry meridian between aggression and sex....In a culture where sex is supposedly all around us and yet true, intense sexuality is in desperately short supply, the show's openness—its willingness to face up to the messiness and potential danger of sex —is glorious and revelatory.[17]

Readings such as these reveal the ways the Buffy/Spike relationship breaks from the representation of sex seen in many media texts in which sex is presented as a sort of ideal space, set outside of time, in which all tensions and conflicts between partners are resolved. As I have written elsewhere, the depiction of "messy" sex displays a kind of feminist consciousness because it "problematizes the conceptions of sex as a context-free, natural act," politicizing sex by revealing the way power relations are constructed and contested within a relationship.[18]

Buffy's relationship with Spike clearly functions on a metaphorical level as well. For example, at this point in the narrative Buffy desires to be dead and her attraction to Spike reflects this: He is both dead and a killer of Slayers.[19] His ability to kill her is clearly part of the appeal for her. Likewise, the series equates physical danger to emotional danger. Beating her in the face in "Smashed," Spike says he doesn't want to hurt her "much." Buffy replies that he hasn't even come close, and he asks her if she's afraid he might. Buffy repeats the idea that Spike cannot "hurt" her at several points during the season, but it is unclear if she means hit her or touch her emotionally. Physical pain is linked to the sexual/emotional connection between the two. Given the frequent assertion that Buffy has "shut down" because of events in her life, particularly her relationship with Angel, the physical violence may be seen as representing Buffy's emotional reawakening.

Similarly, many fans appear to be reading the sexualized violence as a sign of Buffy's power. Fan comments reveal an appreciation of the way "Smashed" and "Wrecked" revealed the extent of Buffy's physical desires. Spike's ability to hit Buffy is seen as indicative that theirs is a relationship of physical and mental equals, while the rough foreplay depicts female sexual aggression in a way rarely seen.[20] It must also be noted that the sixth-season sex scenes provided another role reversal in their presentation of the male body as visual object. While Buffy is covered or clothed in each scene in "Wrecked," for example, Spike is nude. Whatever artistic decisions influence this (it symbolizes the openness of his emotions while Buffy's feelings remain hidden), by the vast number of delighted on-line posts about "nekkid Spike" and the body sock worn by the actor, it is clear that many fans are taking extreme pleasure in the opportunity to gaze at the male body.

The text also complicates the notion of Buffy as sexual victim by questioning her own role in the relationship. Buffy initiates most of the sex, and she is almost always shown on top of Spike. Throughout the season it is quite easy to read Buffy as the sexual aggressor. "Dead Things," for example, parallels Buffy's actions towards Spike with the horrible behavior of Warren, who kidnaps and brainwashes his ex-girlfriend, Katrina. Warren, Jonathon, and Andrew use Katrina like a sex toy, dressing her in a French maid's costume and squabbling over who gets to have sex with her first. When she tries to escape, Warren kills her. Buffy is similarly using Spike for her own sexual gratification without consideration for his feelings. Tricked by Warren into believing that she has killed Katrina, a guilt-ridden Buffy attempts to turn herself into the police. When Spike tries to stop her, she denies that he loves her and goes ballistic, beating him to unconsciousness, telling him "There is nothing good or clean in you. You are dead inside."

She is certainly lashing out at herself in this scene, believing herself to be unclean, but this does not deny the physical reality of what she is doing to Spike. Spike allows the beating, but Buffy is clearly not considering his wishes or feelings. In a sense she is doing the same thing that Spike later does to her in "Seeing Red"—obsessed with her own perspective, she denies the reality of his feelings (which she continues to do all season) and brutalizes him. The parallel is made clear in a dream sequence within the episode, in which Buffy conflates Katrina, whom she believes she has killed, with Spike.

Despite Buffy's frequent abuse of power in her dealings with Spike, there are some rather disturbing implications regarding female desire and female power at play. "Dead Things" and the Buffy/Spike storyline as a whole may be read as depicting Buffy confronting the depths of her sexual desire. Although the sex has been dangerous, we are also told it is the best sex she has ever had (they go five hours straight). Her guilt and horror at the relationship can be read as her being appalled and horrified by her own sexuality, and certainly it is possible to read the text as implying that "good" girls do not have satisfying sex.

In addition, Buffy's power as a Slayer has been linked to her sexuality since season three when Faith makes the link between her Slaying abilities and sexual desire. Faith says that Slaying always makes her horny. Buffy, clearly in denial, replies that it only makes her crave low-fat yogurt. Yet in "Wrecked," Dawn reveals that the first thing Buffy does every night after slaying is head to the refrigerator where "she's such a pig." Willow's magical power was likewise used as a metaphor for lesbian sex. Female power is celebrated throughout the series, yet power is conflated with female desire, which is often dangerous. Faith is unable to control her desires and slips into darkness. Buffy's desire for Angel almost destroyed the world. The sixth season parallels between the Buffy/Spike relationship and Willow's addiction to magic likewise stress the danger of female desire/power.

Buffy finally admits to desiring Spike, but it is a desire that, within the sixth season, at least, she is unable to reconcile. She tells him she cannot love him, and "Seeing Red" makes it clear that this is because he is inherently evil. Her desire for him is thus evil and negative; as she says, the relationship is "killing me." Buffy cannot desire an evil thing without negating her identity as a hero. Her dream in "Dead Things" reflects this death of identity when Katrina's death cry as she is stabbed becomes the sound of Buffy's orgasm as she is penetrated by Spike.

However, Buffy's relationship to Spike, who functions as her shadow, is actually quite crucial in interpreting her status as hero since their sexual relationship dramatizes her attempts at integrating her dark side—a side which she has historically resisted.[21] An important evolution in Buffy's character has been her increasing awareness that the powers of the Slayer are rooted in darkness. Dracula says she is "kindred." Buffy moves from "patrolling"—a duty that she performs as a job—to "hunting." She increasingly begins to recognize the fact that she is the Slayer and enjoys her work—killing. "Fool for Love" carefully establishes Slayers and vampires as symbiotic, linked to each other for self-definition. However, over the years Buffy has resisted her link to darkness. She rejects Dracula's attempts at seduction. In her encounter with the "First Slayer" in a dream in "Restless," Buffy rejects the first Slayer's mandate that she fight alone, with no friends and no love. Buffy tells her, "you're not the source of me," reaffirming her allegiance with the world of the everyday and humanity. Spike certainly represents the seduction of the dark side, repeatedly saying that she belongs in the shadows and darkness with him. Tellingly, when she finally dumps him in "As You Were," she leaves the darkness of his crypt to stride out into the sunlight.

Thus, despite her attraction to darkness, season six dramatizes her rejection of her darker side once again. She rebuffs Spike, as he says, because she would never really "lower herself" enough to be with him. In addition, the fact that it is Spike who is ultimately elevated to a higher level through the gifting of a soul underscores the necessity of his changing. If Spike is to have Buffy, he will have to do it on her terms. Buffy's essential goodness is thus reaffirmed, and she is separated from her dark side yet again.

While Buffy has engaged in some questionable acts over the course of the series, repeatedly she chooses to be "good," preserving her status as hero. However, choosing good seems to involve suppressing and rejecting her sexual desires. It also suggests the lack of integration of those qualities in herself which Spike represents.[22] While season six shows her beginning to integrate those aspects of herself associated with the character of Dawn, Buffy still rejects the "dark side" of female identity. One hopes that the end of the series will depict Buffy's acceptance of these qualities, ultimately revealing her struggle as part of the process on the road to maturity.

As this section reveals, *Buffy* presents some often contradictory attitudes about gender, sexuality, and the body. The series is quite critical of male dominance—the villains of season six are presented as immature boys whose villainy is explicitly tied to their misogyny, while Angel and Spike's journeys as characters are focused around their struggle to move from the dark side—yet the series reaffirms the notion that men are essentially monstrous and that this monstrous masculinity is desirable. While presenting Buffy as physically powerful, her body is situated in a world of male dominance and sexual violence. Buffy's relationship to her own body—her sexuality—plays a key role in the construction of her identity. *Buffy* celebrates female physical power and presents its protagonist as possessing strong desires, but she often experiences her own desire as dangerous. At the same time her body and its desire give her power, they threaten to betray her.

The representation of sexuality in the series suggests that the body is untrustworthy, threatening to erupt out of control and thus in need of taming. The series also reinscribes the sexual body as a necessary component of female identity. However, the complexity and contradictory nature of the text points to the inconsistent ways we often experience the body. In a world in which women gain power from their bodies, even as this power limits and constrains them in other areas, *Buffy*'s paradoxical representation of its protagonist's sexuality strikes me as an extremely "realistic" representation of how women experience their bodies on a daily basis. As Foucault has shown, power neither originates from nor is located in one site but is instead part of a shifting dynamic. The power that one possesses alters, changing or moving depending on one's position relative to others.[23] Thus power, bodies, and identity are always constructed in relationship and as part of a process. The meaning of the body and its relative power shifts as we interact with the world. It is not static, but formed in relation to the outside, material world.

Morality and Relationships

Given the way the narrative stresses relativity and context in its description of female power and desire, it is interesting to examine the ways the series depicts the interaction of its protagonists' bodies with their world. Through its depictions of its protagonists' confrontations with various villains and moral dilemmas, *Buffy* functions as a morality play and asks what actions can/should the physical body take. Over the years, the series has begun to convey a worldview that likewise stresses relativity and the interconnection of all things.

The moral universe of Buffy has shifted over time. The protagonists of *Buffy* have grown up over the years, moving from being high school students out into the "real world." The series' morality and epistemology have reflected the changes and evolution that occur during adolescence. Evolving from a worldview in which the distinction between right and wrong, good and evil was clear, the series has demonstrated an increasing complexity in the moral dilemmas encountered by its protagonists. In many ways, Buffy's increasing moral complexity echoes the process that Carol Gilligan has described as an ethics of care.

Gilligan writes that moral reasoning for women moves through three perspectives, moving from "an initial concern with survival to a focus on goodness and finally to a reflective understanding of care as the most adequate guide to the resolution of conflicts in human relationships."[24] Each perspective represents "a more complex understanding of the relationship between self and other and each transition involving a critical reinterpretation of the conflict between selfishness and responsibility."[25]

While the usefulness of considering this moral development as specifically gendered one is open to debate, it is interesting to consider how the show's "adult" morality in later seasons reflects a moral reasoning based on care and an understanding of the connection between things. As Rhonda Wilcox has observed, *Buffy* has always presented the necessity of connection to friends and family.[26] However, the series has moved from presenting moral dilemmas as external, with solutions based on rules and absolute standards of judgment, to a more complex view in which characters learn to question their own selves and actions and in which they increasingly need to come to terms with the effect of their actions on the world. The series presents all characters as possessing the power of choice and action and needing to accept the responsibility that comes with this power.

It is important to note that for its first two seasons *Buffy* featured a world that is actually rather orderly and governed by a set of dependable rules. The world of *Buffy* is a dangerous space, but it can be navigated safely, once one has access to the right tools and knowledge. For example, the rules of conduct for mystical beings are certain. Vampires cannot enter houses unless they are invited. They hate holy water and crucifixes and can be killed by stakes or decapitation. There are always spells or incantations that can be found through research in Giles' library to stop the evil, or the monsters can be physically destroyed. While each episode starts with a mystery, it eventually becomes clear to Buffy what is going on. Usually she solves the dilemma by accomplishing a set task, either magical or physical.

For example, in one attempt to break free from his imprisonment behind a mystical barrier, The Master sends out the "Chosen One" who has the ability to suck souls and transfer their power to The Master. To stop The Master, Buffy simply kills the Chosen One. A demon raised by Spike and Angel, who according to the ancient scrolls cannot be destroyed "by any weapon forged," is defeated after Xander realizes that the scroll was written before high-tech weapons; Buffy blows him up with a rocket launcher.

Likewise, although the monsters hide their identities and appear to be human, their true nature is always eventually revealed. Before a vampire can bite its victim, it must show its true monster face. Unlike Dracula-like vampires, who continue to resemble humans except for their teeth, the vampires in *Buffy* are clearly Other, clearly inhuman. Whedon has said that he made his vampires explode into dust when staked because "it shows they're monsters. I didn't really want to have a high school girl killing people."[27]

The monstrousness of the vampires maintains a distinction between good humans and evil monsters. This means that there is little threat in watching them die and removes any blame or guilt on the part of the protagonist who slays them. It means Buffy's use of force never threatens to turn back on her. Thus, for the first few years there was very little relativism, no blurring between right and wrong in the series' overall worldview in which morality is certain. Importantly, these distinctions are easy to maintain in earlier seasons because the threats Buffy confronts are all external.

However, later seasons complicate the certainty of actions as moral dilemmas and character actions become more complex. The most notable example of this is reflected in the breakdown in distinctions between human and monsters. Spike, an out-and-out villain, increasingly reveals his ability to do good. He makes ethical choices, risking his life to protect Dawn because he "made a promise to a lady" and repeatedly revealing the depth of his human, caring emotions. Humans likewise increasingly do bad things. In "The Gift," for example, Buffy has spared the life of Ben, a human who turns into the evil god Glory. She cannot bring herself to kill a human, even though she knows it means that Glory is likely to return. Giles tells Ben, "Buffy…couldn't take a human life. She's a hero you see. She's not like us," before smothering him. Giles' assertion that Buffy could never kill a human is actually contradicted by Buffy's own statement earlier in the episode when she tells her friends that she would kill anyone among them who attempted to kill Dawn in order to save the world. Even more notable is Willow's sixth-season torture and murder of Warren by flaying him alive and her subsequent attempt to destroy the world. Although it is suggested that dark magic has overwhelmed her, her actions are nonetheless as villainous and horrific as those of any of the various monsters Buffy has fought over the years.

Later seasons, most particularly season six, break from the pattern of previous years in several ways. Most notably, the monsters of season six are internal; psychic dilemmas and interpersonal conflict are the focus of most episodes. It is also notable that this is the first season where Buffy has not physically confronted a main villain. There are no set external physical tasks which Buffy can accomplish to put things right. Instead, it is clear that the only way for characters to save themselves and consequently save the world is to understand themselves, their own actions, and the effect of their actions on others.

For example, Buffy's depression after being brought back from the dead has resulted in her isolation from all the characters, particularly her sister Dawn. She is so intent on protecting Dawn (who represents Buffy's own innocence) because of her own fear of life that she can not see how her constant protection and babying of Dawn are crushing Dawn's own development and sense of self. Xander's fears that he will repeat the life of his father have led him to dump his fiancée at the altar; he moans about his inability to do anything right, not seeing that he has control over his own actions and can choose not to be like his father. Most notably, Willow is so grief-struck by the death of Tara that she is unable to care about other people, including her friends. When Giles doses Willow with the true essence of magic, it allows her to connect back to her human emotions again, which include feeling the pain of the world. Willow's immediate reaction is to end the world's pain by destroying the earth. Buffy, Willow, and Xander are stagnating because they are unable to see how their actions affect others.[28]

In the concluding moments of the season, Buffy is trapped in a pit with Dawn. Buffy's status as hero is rendered useless: External power plays no role in the heroic resolution; it is just plain old human love and emotion that reconcile all of the characters with life. Xander ends up saving the day by reminding Willow of his unconditional love for her, placing her in connection once more with the world. Buffy realizes that Dawn has an independent identity and that she cannot protect her forever, at the same time reaffirming her own desire to live.

Buffy's take on morality is particularly interesting to consider in terms of what it has to say about women's power. Elyce Rae Helford has written that *Buffy* prescribes a certain kind of anger management for its heroines, with Buffy repressing her anger in the way of good middle-class girls.[29] Certainly, it is easy to read the series as being cautionary about women's power, as for example, Buffy's out-of-control sexuality and Willow's sixth season storyline, in which her power is depicted as being out of control.[30] However, what can be read as repression on one level also reflects a distinctive attitude regarding the proper way to handle power, which once again stresses responsibility and care for others.

In *Buffy*, it is women who hold power: Buffy, Kendra, and Faith are Slayers; Willow and Tara are witches; Dawn is the Key. Buffy, Faith, and Willow's powers and use of their power have been central to the storylines. Their storylines help articulate what a hero *should* be. Both Faith and Willow come to grief because of the way they handle the responsibility of their power. In contrast, Buffy, as mentioned in an earlier section of this chapter, uses her powers for others and to protect social stability, and thus her use of force never threatens to turn back on her.

Faith is a Slayer from the "wrong side of the tracks," who is called after the death of Kendra. As Wilcox has noted, Faith's troubles stem from her isolation from community[31]; however, her storyline also demonstrates the seduction of power and the dangers of living by one's id. For Faith, slaying is fun: It gives her a sexual charge. She believes the power she possesses as a Slayer makes her better than others. Like Spike, Faith functions as Buffy's shadow, representing that side of Buffy that she is desperately trying to suppress.[32] For example, Faith tries to get her to admit that slaying turns her on. Buffy feels the pull of this dark side—she follows Faith's lead when she begins to act irresponsibly with her powers. The two girls use them to break and enter into a weapons store in "Bad Girls." Buffy finally rejects impulsive, feeling-centered action after Faith mistakenly kills the mayor's aid, a human. She sees that Faith's ways can lead to chaos. Faith is presented as following her id—her actions are motivated by pure emotion—jealousy of Buffy, desire to posses the things that Buffy has, like love, which the Mayor offers her. Following her id without care for others leads to Faith's destruction. She has power but uses it irresponsibly.

This theme is echoed three years later. For many years Willow seemed an unlikely candidate to be the Big Bad. But many aspects of her character which come to the fore in season six have been in play since the early seasons. These include her impatience with suffering, her desire to escape from her past identity as a high school loser, and an essential arrogance over the use of magic which she sees as functioning like computers. Willow's power, which steadily evolves over the years, comes to fruition in the season-five finale, "The Gift," in which she is described as Buffy's "big gun," the only one of them with the power to face off against the goddess Glory. Her power is central in season six, when she possesses the power to bring Buffy back from the dead.

Buffy's resurrection is, in fact, irresponsible. It is "against all the rules," as Tara says, and Willow has not informed the Scoobies of all the risks inherent in her plan to return Buffy from the dead. The resurrection is done without considering the consequences. It turns out Buffy was at peace in Heaven; Willow has returned her to what Buffy sees as a hell on earth. But perhaps the most important consequence is how the resurrection affects Willow, setting the path for her downward spiral.

Willow's great fault is her arrogance about her power. When Giles chastises her for the resurrection because the spell could have had dire consequences, not only for her friends and the spell's effect on Buffy but for the world, Willow's simultaneous arrogance and desire for recognition are apparent. She tells him, "I thought you'd be impressed...I was amazing...I did what nobody else *could* do." Giles tells her, "There are others on this earth who can do what you did. You just don't want to meet them." Willow replies, "They're the bad guys. I'm not a bad guy." Willow wants to separate identity from action. She is unable to see her actions as a problem because her sense of herself as "good" means that her actions must be.

Willow's arrogance about power is clearly linked to her own insecurity about who she used to be. Her finale rampage is presaged when the shy child inside comes out in "Seeing Red" when she learns about Buffy's secret affair with Spike. She is terrified that she has been left out of the loop ("Did everybody else know? Am I the only one?"). Her resentments about being the sidekick are articulated during the finale as she faces off against Buffy and Giles. While it has certainly been used to help others over the years, it is clear that Willow sees her power as an escape from herself.

Willow's problems with magic in season six are tied to her historical use of magic as a quick fix. In the fourth season, for example, Oz's leaving devastates her. While Buffy tells her things will get better, she is unable to wait and tries to cast a spell to fix things. The spell, of course, goes dangerously awry. Willow seems chastened but continues the pattern. In season six, she

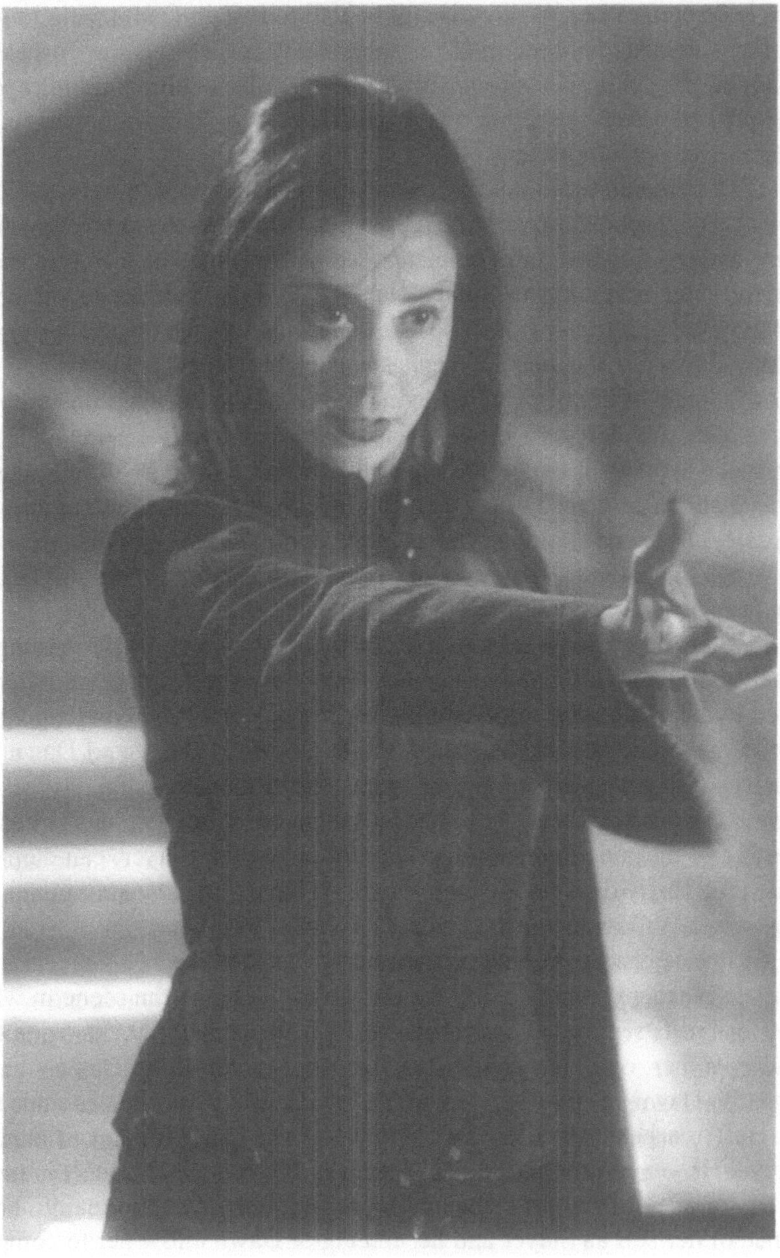

Figure 8. Evil Willow.

ends a fight with Tara by casting a spell of forgetfulness. When Tara discovers this and rebukes her, Willow is still unable to refrain from trying to get a quick fix; she casts another spell to make Buffy forget she was in heaven. Once again, the spell goes wrong. Willow essentially commits mind-rape on the people she loves most, but she is unable to see from their perspective how assaultive her actions are.

Buffy's personal problems in season six are also linked to her unwillingness to accept responsibility. At the beginning of the season she yearns to be back in heaven, finished and complete, free of her duties as a Slayer, sister, and friend. Her attraction to Spike is depicted as her attempt to dissociate from the world and her responsibilities. Consequently, she begins to act irresponsibly, ignoring the pain of her sister and her friends. In "Gone," she takes advantage of being invisible to escape from the normal constraints of her life. She uses her invisibility to harass the residents of Sunnydale and gets a social worker assigned to help Dawn in trouble. In "Normal Again," she wants so desperately to be free of her life as it is that she imagines that Sunnydale, her friends, and her life as a Slayer are dreams and that she is actually a mental patient. She seeks to escape into this alternate universe by killing her friends.

While the entire sixth season, like "Bad Girls," shows Buffy verging towards the dark side, ultimately she comes back to herself. Importantly, she is restored by her accepting responsibility for her power by acknowledging her ability to make choices. In the season finale, "Grave," Buffy and Dawn find themselves trapped in a pit fighting against earth monsters that Willow has summoned to destroy them. Anya has told them of Willow's plans to destroy the earth and that no supernatural power, not even the Slayer, can stop her. Despite this, Buffy does not give up, saying "I have to try. I'm not gonna just sit here while Willow incinerates what I'm chosen to protect."

This line is crucial in her reclamation of her identity as a Slayer as she chooses to reaccept responsibility for the world. It is also the scene in which Buffy, for the first time, allows Dawn to fight for herself. At the same time she accepts her own responsibility, she stops controlling Dawn—caring enough for Dawn to give her freedom to make her own choices and decisions. Buffy earlier told Giles that she felt like she had left part of herself in the grave. However, among the final scenes of the episode, is Buffy crawling from the "grave" of the pit bringing Dawn with her. Metaphorically her acceptance of her self as Slayer and her freeing of Dawn allow her to complete her rebirth.

Overall the series repeatedly stresses the interconnection of things. Buffy is continually netted within the web of community. Likewise, power, we gradually learn over the years, is a double-edged sword that needs to be used responsibly. Part of what makes the use of power responsible is a consideration of the consequences—particularly how power will affect others. This morality reflects a particular philosophy towards the body. While it does not suggest that an ethics of care is part of some essential female quality, it certainly breaks away from having abstract absolutes as the guide to action in the later seasons. Characters repeatedly learn that rights and wrongs are not clearly delineated, and situational context is important. It stresses the idea that moral choices, like all actions, have consequences for the self and others. Physical occurrences in the real world may follow as the result of a choice. The self is always as netted within a web or relations, shaping and shaped by the world around it. In this way, the body, the self, is not a discrete entity but is part of a system.

A guiding theme of *Buffy* is its focus on the consequences of action. *Buffy*, like male-centered texts, thus shows the subject's power to shape her world, yet the series' focus on consequences and the construction of an identity formed through a web of relations means that as Buffy acts she is also acted upon. While male-centered texts construct a heroic subjectivity based on isolation and denial of the body, *Buffy*'s emphasis on relationships and sensation creates a world in which subjectivity is formed through the sensation of existing within the material world and which is clearly formed in relation to that world.

Conclusion

Buffy has been widely interpreted as a feminist show because of the way the series deconstructs the notion of femininity as exclusive of female power.[33] Others have argued that the "feminism" that *Buffy* supports is most consistent with a liberal feminism focused on white, middle-class girls.[34] The series has been criticized for its support of capitalism, its depiction of race, and until recently, its insistence on heterosexuality.[35] This analysis likewise points to some of the problematic and contradictory ways the series represents gender issues, most particularly, its take on male and female relations and the link between female power and dangerous desire. However, Buffy's status as a feminist text ought to be considered alongside the way the series alters existing models of the action text.

As discussed in Chapter Two, the traditional male action text presents the body as a means to an end. Although action texts are tales told through the body, most often we actually see a denial of the body, of materiality, taking place. Male heroes' ability to master and dominate their bodies functions as a display of ephemeral qualities such as their spirit, drive, and willpower. Male-centered texts reveal these qualities, not the body, to be "the self," and their heroic subjectivity is thus one that exists outside the body. If men's bodies are put on display to show their ultimate escape from the confines of the body, in contrast, such an outlet has not been traditionally available to women in action stories. Indeed, in the case of the hapless women in need of rescue in most texts, the helplessness of their bodies is significant. Their physical weakness implies that they cannot rescue themselves, not merely because they are physically weak but, in fact, because they can not get past their bodies. Such women are effectively victimized *by* their bodies.

Although *Buffy* was consciously created as a response to this image of women victimized through their inability to escape their bodies, the series actually reinscribes the centrality of the body in various ways. Textually, the series encourages viewers to experience the "emotional realism" of the situations in which its characters find themselves. Watching the series allows viewers to experience an almost physical relationship to the text, while recalling and reliving their own embodied experiences. While most action texts invite the viewer to experience the visceral thrill of stunts, *Buffy*, like *Nikita*, encourages its audience to experience and re-experience emotions, a move which works to value everyday human existence in the material world.

Buffy's depiction of the sexual behavior of its hero is frequently problematic but points to the often contradictory ways we experience the body in the world—as both a means of power and a threat. It is significant that the world in which Buffy lives is *not* safe. Even trusted friends can be taken over, and the "truth" about them may not be realized until too late. Sexuality can be a dangerous thing, changing an otherwise wonderful guy into a raging demon. Buffy and other girls figure out how to identify and control hidden evils like their unruly boyfriends. In this way, *Buffy* normalizes an environment of sexual danger at the same time it functions as a call to arms for girls to stand up for themselves in an era in which traditional notions of female passivity are no longer feasible.

Despite presenting its heroine in a dangerous environment, Buffy stresses its protagonist's agency. While *Nikita* depicts a hero who is primarily acted upon, and *Aeon Flux* deconstructs the notion of heroism by revealing the futility of action, a guiding theme of *Buffy* is its focus on the consequences of action. *Buffy*, like male-centered texts, thus shows the subject's power to affect her world; yet the series' focus on consequences and the construction of an identity formed through a web of relations means that as Buffy acts she is also acted upon. This view acknowledges the ways that we are often constrained in the world but at the same time acknowledges the power of individual action.

Buffy contrasts to male-centered texts in which the hero acts upon the world. Although his body is frequently acted upon by external forces, the hero's "true self" is shown as being located in the realm of mind and spirit and is untouchable. His subjectivity is closed, already formed, independent of the experience of his body or the world around him. In contrast, *Buffy* suggests the ways that identity is constructed through process, relationship, and the physical sensations and experiences, both positive and negative, of being in the world. Whereas in traditional texts the body is a means to an end, in *Buffy* the experiences of the body is the end itself, a move which works to valorize the "feminine" realm of the body. Heroic transcendence for Buffy is inextricably linked to a sense of bodily continuance.

[illegible]

CHAPTER SIX

Female Heroes and the Body in Postmodern Culture

This analysis is based on an understanding of the "body" as both a product and a process, a way of knowing and marking the world as well as a material object.[1] Like most representations of females in mainstream media, the bodies under discussion in this text are white, thin, frequently sexualized by camera work and narrative, and meet contemporary standards of female beauty. However, this is not to say that they offer no disruptions from traditional representations of female bodies and female power. Much of this analysis has considered the ways in which textual strategies of representation, such as narrative, camera work, or dialogue construct particular ways of understanding the body that are markedly different from male-centered texts. In contrast to the suppression of the body occurring in most male-centered texts, these three series reinscribe the centrality of the body. *Nikita* presents the oppressed body as the central problem its protagonist confronts at the same time as it celebrates the sensation of the body in the world. *Buffy*'s narrative and other textual features reveal the series' central concern in asking what it means to be embodied in the world, an approach which underscores both the dangers and pleasures of embodiment. *Aeon Flux* deconstructs traditional ideas of heroism and critiques rationality through its presentation of the grotesque, disruptive female body.

It is important to observe that the body suppression of most male-centered texts fits in amongst a long philosophical tradition of Cartesian dualism in which the dominant image of human beings is "one of control, domination and sovereignty."[2] Bryan Turner has written that "Cartesianism preeminently expresses the virtues and values of possessive individualism, of the rational enquiring mind and mastery over the natural environment."[3] Importantly, Cartesian ideology emphasizes asceticism, control and production, qualities required by the cultural demands of early capitalism.[4]

However, with the changes of late capitalism we have moved into what has been described as the "postmodern" era. We live in a postindustrialist culture that is part of a global economy and that is dominated by consumerism, fashion, advertising, and new leisure industries. The result is a cultural contradiction "between hedonism in the arena of consumption and the remaining legacy of asceticism in the labour process."[5] In other words, we need to be thrifty, self-controlled, and disciplined in order to make products, but we are encouraged to abandon discipline and self-control and to satisfy all our desires when it comes to purchasing goods. Contemporary ideology promotes living for the moment, self-expression, pleasure-seeking, and style.[6] Correspondingly, the postmodern view of the human being is of an "anxious, thus playful creature in search of personal satisfaction through new forms of intimacy and personal experiences." [7] Importantly, postmodern theory critiques the Cartesian view of reality, emphasizing social constructionism, and knowledge and belief as relative and dependent upon perspective.[8] Turner observes that the postmodern age, rather than doing one away with the tradition of individuality, has created a new kind of individuality based on "emotionality, sensibility and sensuality."[9]

The representations of the body in female-centered texts mediate contemporary cultural concerns in different ways. In addition to emphasizing intimacy, sensuality, personal experience, irrationality, the social construction of identity, and relativistic thinking, these series present the body as a borderland—a particularly potent cultural symbol at this time. This chapter ties the representation of the body in these series to critical observations about the postmodern body and historical understandings of the female body to offer suggestions as to why the new model of hero is female. It discusses how these series through their "feminine" strategies and concern with embodiment address contemporary cultural concerns over mortality and connect to larger cultural trends regarding community, individuality, and identity.

The centrality of the body in the three series examined here alters the subject/object relations of most male-centered texts in which the dominance of the mind over the body is paralleled by the hero's domination of others. Female heroes have a very different relationship to the external world, to others, and, consequently, to their own bodies than male heroes. While male heroes frequently function as boundary-crossing figures existing in a borderland, the body of the female hero *is* the borderland, positioned as both vulnerable to penetration by and reactive to her external environment.

As with action heroes, the environments in all three series are hostile. Female heroes fight to maintain their bodily integrity against invasion by external forces. Buffy lives over a literal Hellmouth and is in constant danger from death, possession, and penetration—not only physically, but emotionally. Nikita, a spy, must constantly find a way to "live" despite facing the death-focused environment and her deadly job and despite having had her identity and her privacy stolen from her. Aeon lives in a world of permeable boundaries, which is completely unstable. These female heroes are never safe, are forced to navigate dangerous terrain, whether the danger lies in physical terrain, physical actions, or emotional or interpersonal terrain.

Male heroes suffer physically because of the way their identities are constructed as existing outside their bodies; their injuries are flesh wounds, never reaching beyond the level of the skin. In contrast, the injuries experienced by Nikita and Buffy, although less physical than psychic, permeate their entire beings. Aeon Flux is equated to the crumbling, disintegrating landscape that surrounds her; her own actions destroy her. The heroes in this study both begin and end the story *already suffering*. The female hero does not encounter suffering, as do the male stars, moving from a position of equilibrium to danger and then fight and bleed to finally restore equilibrium. Instead, she starts in disequilibrium, and throughout the text she only succeeds in maintaining the status quo, maintaining disequilibrium. Nikita remains trapped inside of Section; Buffy defeats a range of villains only to die and be brought back to fight again. Aeon endlessly repeats the same actions. There is no ultimate triumphant ending in which the hero gains control over the external world, like that experienced by Rambo.

Obviously, this is partly because of the serialized nature of television. However, female heroes in films experience a similar continuance of suffering. As Kirsten Marthe Lentz found in "The Popular Pleasures of Female Rage," the victimization experienced by male cinematic heroes exists because "the audience narratively understands that pay-back is near and that the painful interlude is a temporary form of victimization....However, for the [female] hero...the dynamic reverses so as to create the sense that it is her agency/freedom which is only temporary and that she will soon return to her normally victimized, compromised position."[10] Likewise Sarah Connor in *The Terminator* heads off into a sunset with the awareness that "a storm" is coming—the future wars with machines. In *Alien 3*, Ripley's body is invaded, impregnated by the alien. In the fourth film in *Alien* series, her body, killed in the third film, is reproduced against her will by scientists. Her new body contains the DNA of the alien. Her plundered body is pathologized, infected by the evil Other.

It is vital to observe that postmodern theories of the body construct an image of the body that is similarly threatened by invasion and decay. Anne Balsamo explains Arthur Kroker's discussion of the body and contemporary culture in "Body Invaders." Kroker connects conceptions of the debased body with contemporary cultural issues like AIDS, addiction, and anorexia.

Postmodern definitions of the body are thus representative of "social anxieties concerning control and safety."[11] Cultural fears of conspiracy, alien abductions, and technology may similarly be read as expressing the fear of a body under invasion. John Cawelti and Bruce Rosenberg likewise observe that recent Western films reflect a distrust in the system of law and justice and sees the popularity of spy novels as reflecting a contemporary public discomfort with the possibility of clandestine organizations abusing their power.[12]

Currently, there are a wide range of television series which display themes of the threatened body. For example, *Strange World*, *The Pretender*, *The X-Files, Millennium*, and *The Net*, all display themes of surveillance, conspiracy, and bodies living under constant threat of invasion. The hero of *Strange World* has a virus placed in his body to ensure his cooperation. *The X-Files, Dark Skies,* and *First Wave* are riddled with the threat of an alien takeover. Max, the heroine of *Dark Angel*, was bred in captivity by a secret government organization that continues to hunt her. *G vs E* is about the conspiracy of Morlocks, those who have bargained with the devil, to take over the world. In *Poltergeist: The Legacy*, the hero and her daughter fight evil forces that erupt into the normal world from the other side. *The X-Files'* hero, Scully, has a metallic object implanted in her neck by aliens; her partner Mulder's own mind is suspect; he cannot trust his own memories about his family's past; the series' motto is "trust no one." A conspiracy threatening the end of the world plagues the heroes of *Millennium* and *First Wave*, while the hero of *The Net* finds herself on the run from a conspiracy who have stolen her identity through computer technology. Sam, the hero of *Profiler*, is continuously stalked by an obsessed serial murderer who kills everyone close to her in his attempt to possess her.

However, given the fact that these embattled heroic bodies are everywhere, it is necessary to ask why *most* new action series stars are females. Even series not revolving around a single woman still feature women, as in *The X-Files*. Why is it that in the era between 1960 and 1990, one could count the number of female heroes on one's fingers, but since 1992 there has been an explosion of such images? I believe that historical views of the female body may be a particularly expressive cultural symbol at this particular time, representing both a sense of constant victimization and the possibility of persevering. At the same time, women suggest a different response to the demands of society than previous father-figure heroes.

Importantly, Kroker argues that "women's bodies have always been postmodern" in that the female body functions as the sign of "the body debased, humiliated and inscribed to excess by all the signs of consumer culture."[13] In other words, the images of embattled women, who are without a doubt *heroes*, appeal to the larger television audience, perhaps because fighting female bodies express more poignantly the cultural fear of a body under invasion. The female body, historically coded as for-now-and-ever passive and victimized, is perhaps *more* effective at expressing a totalizing view of oppressive social forces which constrict the individual. In *Men, Women, and Chainsaws*, Carol Clover has similarly argued that the female hero of slasher films appeals to the masochism of the primarily male audience.[14] However, unlike the final girl of slasher films, there is no "triumphant self-rescue" for these heroes, at least not one that lasts. Thus, the female body signals that there will not be an escape, further aiding the sense of claustrophobia created by such films.

However, I do not want to equate the threatened postmodern body of *The X-Files'* Fox Mulder, for example, to the series under investigation here. It is important to note that, like most postmodern theorists, the writers, producers, and directors of each of these shows are overwhelmingly male as are the corporate executives who purchase and broadcast these images. Likewise, in reflecting upon the writings of the postmodernists, Balsamo observes that the "compulsion" of such theorists, overwhelmingly male, to "theorize the condition of postmodern bodies as something new or even exhausted is due to the novelty the male body experiences coming under this totalizing system for the first time; understandably, panic results."[15]

She argues, and I agree, that this position renders gender meaningless, reduces it to the level of discourse while losing all reference to the materiality and historical presence of the body. Balsamo observes, "Such is the fate of the female body in the postmodern cultural imaginary: an always silent/silenced conceptual place holder in hysterical male discourse. As before, her excessiveness threatens the very order to the system."[16] Thus, it is necessary to address how these threatened heroic bodies are specifically gendered and shaded by other discourses that change the meaning of their embattled state.

At the same time postmodernism reflects a concern about threats to the body from the external world, it also regurgitates ancient fears about the uncontrollable internal processes of the body, in other words, a fear of the feminine. As noted earlier, the female body has traditionally been tied to a grotesque, carnivalesque sensibility. Bakhtin has written that "woman is essentially related to the material body lower stratum; she is the incarnation of this stratum that degrades and regenerates simultaneously."[17] The boundaries of the female body are flexible, permeable, more easily threatened with invasion than the tightly contained male body; *at the same time* the very liminality of the female body is threatening, evoking anxieties about mortality. It is the female body, the maternal body, who "through menstruation, pregnancy, childbirth, and lactation, participates uniquely in the carnivalesque drama of 'becoming,' of inside-out and outside-in, death-in-life and life-in-death."[18] Thus, the inner workings of the (female) body are not only threatened but contain the seeds of power. Because of this, I think the female body offers a possibility of transcendence that is rarely discussed by male theorists. It is important to examine not only the fact that the heroes exist on a border state but *how* the heroes negotiate these dangers.

As previously observed, action heroes traditionally reflect the dualist concern with the mind's control over the body. The hero displays his worthiness, his uniqueness, by having a particularly pure spirit and strong mind that allows him to control his unruly body. By conquering his (feminine) body, the hero conveys a sense of transcendence, of individual immortality. From classic Westerns to *Rambo* and *Die Hard*, action texts may be littered with corpses, but the hero survives against all external forces that oppose him, because *he*, his mind, his will, his spirit, is unique. His continued life represses the fact of and inevitability of death, simultaneously and paradoxically suggesting an immortal soul.

Obviously the series under observation also repress the fact of death for their heroes. However, they do so differently. Both Buffy and Nikita hate the path their lives have taken but are forced to live with and recognize the limitation of their bodies and their inability to physically alter their external environment. In other words, both seek to find a way to live in the face of their continued conscious confrontation with death. This pattern is also seen in series such as *The Profiler* and *Xena*. Xena constantly struggles with her bloody past and the present necessity to battle evil without being overwhelmed by it herself; Sam struggles to trap serial killers and live happily with her daughter while being stalked by the murderer Jack. *Aeon Flux* similarly reflects the ever-present awareness of death and the inability of human action to alter the reality of death's presence.

It is of critical importance that in each of these series, most of the narratives focus less on the physical prowess of the hero—or even her intelligence—but the lived *experiences* of her body. While the classic hero works to control the body—in effect, to leave the body behind—female heroes ground their struggles within the body. Self-transcendence for them is inextricably linked to a sense of both spiritual *and* bodily continuance.

One way the series emphasize the primacy of the body is through the use of romance and other "feminine" forms of presentation. It is important to observe that, as in romances, bodily senses in the three series are as much a source of pleasure as they are of danger and pain. For example, Nikita loves her oppressor, Michael, just as Buffy and Aeon love their archenemies. Is the experience of falling in love one of danger or one of bliss? The fact that there is such uncertainty about the meaning of sensual moments both within life and within the texts transforms the hero's relationship to her body. If the body offers a route to pleasure (and power over the lover) as well as danger, you do not want to quash it, deny it. Thus, the body is not to be conquered and controlled but to be lived with.

As earlier observed by Janice Radway and Tania Modleski, romances often contain social critiques which reveal the dangers of patriarchy, particularly the lack of domestic safety. In romances, the body of the hero is the instrument through which she learns to mediate and navigate the dangers of female existence. However, it is also her primary source of identity and pleasure. The romantic elements of *Buffy* and *Nikita* work similarly. Indeed, I think that the most interesting element in these shows is the way they so strongly blend romance with action. Action sequences and fighting work to make explicit the life or death themes that exist but tend to be hidden within the center of the romance narrative. For example, many of the monster metaphors of Buffy reflect the hidden, unspoken dangers of sexual relationships.

Just as romances critique male violence, they also promote a mindset that valorizes a "feminine" ethic of caring and cooperation. *Nikita* clearly criticizes "masculinist" epistemologies that value domination. Tom Carson has called *Nikita* "La Femme Dilbert" because of Section's oppressed workers under the control and surveillance of their superiors, equating the atmosphere in Section to modern-day corporate workplaces.[19] Though Carson does not say this, it is interesting to observe the patriarchal structure of such work settings. The silence that lies behind the penetrating gazes of the Section bosses reveals their dominance over their workers. Section's surveillance culture disciplines the spies; they not only learn but eventually embody the values of their bosses. The workers of Section are thus indoctrinated with notions of dominance, invested with the drive to succeed as individuals, and, like Michael, replicate the very attitudes that oppress them. *Aeon Flux* similarly pokes fun at Trevor's desire to control others, while Buffy focuses on the consequences of one's actions on others.

At the same time, *Buffy* and *Nikita* also rely on feminine stereotypes to justify their use of violence. These heroes are under perpetual threat by those in their personal/domestic spheres. Kirsten Marthe Lentz has noted that the cultural discourses currently surrounding women and guns present women as victims of inevitable male violence, thus necessitating the use of weapons and female violence.[20] Similarly, Nikita must be violent or she will be killed by her bosses, and Buffy needs to fight or vampires will eat her. Additionally, female strength in *Buffy* and *Nikita*—as well as *Xena*, *Profiler*, *Poltergeist: The Legacy*, and *Voyager*—tends to be maternal in that both use their skills to protect weaker individuals like children.

Unlike the classic hero, who, as Tompkins observes, tends to be concerned with controlling others, the female hero seeks to protect her body in an environment which seeks to remove her self-determination. However, we see that at the same time Nikita and Buffy are individuals fighting external forces; they are also framed by their connection to others. The shows therefore endorse the "feminine"—connection to others, caring, emotions, the experiences of the body. While perhaps promoting an essentialist feminist perspective, this perspective is also relatively revolutionary when viewed against the context of a culture, not to mention a genre, which has tended to repress the feminine.

The ultimate value of this perspective might be found in the viewer's reaction to it. Susan Douglas notes that the popularity among young girls for groups like the Spice Girls and Madonna was due to the fact that they addressed the contradictions that young girls experience in contemporary culture. As Douglas observes,

> They are told that if they aren't sexy, they are nothing, but that their sexuality is dangerous to them and threatening to much of the rest of society. Boys are supposed to want them, but they aren't supposed to want boys back—thirty years after the women's movement began, the term "slut" is alive and well. The loss of self-esteem that all too many teenage girls suffer when they learn to silence themselves, to censor their real desires...has kept Mary Pipher's *Reviving Ophelia* (advice on how to help girls survive adolescence in a "girl" hostile environment) on the best seller list for more than two years. The popularity of the Spice Girls is due to the fact that they tell girls that these contradictions can be managed.[21]

Douglas also comments that the scorn with which such groups are received by critics is part of devaluing the feminine. The negative effects of denying the feminine parts within the self have been well documented. However, as she says, the appeal and importance of cultural forms which value the feminine cannot be overlooked.

One reason this valorization of the female is so important is that it seems to signal a different way of thinking about questions of identity and individuality than has previously inflected most mainstream television shows as well as most action texts. For example, *Buffy* and *Nikita* both utilize major techniques drawn from or related to "feminine" culture such as the codes of romance, soaps, dance music, and the importance of emotion. As John Fiske has noted, "feminine" texts tend to be "open texts." The serialized nature of both series, with story lines continuing over time, means that relationships develop and change. Relationships, ongoing and within individual stories, are always more important in the narrative and visual techniques than the supposed action sequences or individual mission of each episode. Conflicts caused by the villains are paralleled, and often overshadowed, by the interpersonal conflicts existing among the cast.

In this way, the importance of relationships within the narrative are boosted to the level formerly occupied by the conflict with the villain. In effect, the series are no longer about the conflict between the ideological positions of the "good" hero and "evil" villain, but are instead *about* conflicts in relationships. Because of this, resolution is much more difficult to achieve than in traditional action texts. Thus, instead of implying that one can discover a final, absolute truth, the series offer a more postmodern epistemology in which knowledge of the truth is dependent upon perspective.

Moreover, as in soaps, *Nikita* and *Buffy* appear addicted to analyzing emotion, addicted to the intensely pleasurable moment of revelations, but the revelations are always temporary and fluid; the tension remains, and numerous high points of emotion are able to stay in play. In sustaining the enigma of the male other, *Nikita* particularly plays with romance narrative but actually shies away from the happily-ever-after resolution which reads as a recuperation of the female hero into the patriarchal order.

It is vital to recognize that this is not the same kind of sexual tension that is seen in many "romantic" investigative TV shows such as *Remington Steel* or *Moonlighting*, in which the romance was endlessly deferred, and each week saw the characters at essentially the same place in their relationship. Instead, as in soap operas, *Nikita* and *Buffy* feature a definite progression *forward* for each of the characters. Contrasting the first season to later seasons of each program, it is clear that there is a different relationship existing between Nikita and Michael and between Buffy and her friends. There is more knowledge about each other, but at the same time, there is more suspicion. As in soap opera viewing, a sense of history emerges in both *Buffy* and *Nikita*, and the interpretation of the text becomes highly dependent upon the viewers' previous knowledge of the characters and their motivations.

Therefore, Buffy and Nikita are essentially constructed as selves-in-relation, dependent upon others and the tactile operations of daily life for self-definition. Again, this can be seen as a way of restricting and containing the feminine, returning the female hero to the private domestic sphere. However, I think something else is also occurring. While being a self-in-relation is viewed as a weakness by the male hero since it suggests the need for another, what the female hero (and many women) understand is that it is also a way of extending the self through communicating with others and taking up far more space than can be contained in a single body. Likewise, the continuity of life is far greater than the span of an individual's solitary existence.

The valuing of relationships, of lived bodily experiences that are communicated to others, works to valorize feminine culture and also connects to an important change in regards to culture and mass communications.

As Catharine Lumby has observed,

> The very idea and structure of the modern public sphere relies on the exclusion of women and all things associated with them....The family, the body, sexuality and relationship issues are relegated to the "feminised" private sphere while political and economic interests are located in the "cultural" "masculine," public domain...this opposition provides self-justifying ground for the exclusion of everything aligned with the feminine from the public sphere.[22]

Thus, Lumby argues that the contemporary debates about the "contamination of traditional news-values by the irrational, indecently personal focus of tabloid journalism emerges as a desire to police the boundaries of a symbolic masculine space."[23] One must also note that "traditional news values" reflect not only a masculine viewpoint but a westernized, white, middle-class attitude about the need to separate the individual self from community.

Crucially, one difference between traditional news reporting and the tabloids is the way they represent the human body. Traditional news reports, like the traditional action hero, frame the body as something to be enhanced or cured, the "object of expert knowledge" or one which needs "institutional control."[24] Tabloid media, on the other hand, are "grotesque" and feminine. Tabloids construct a lived body, connected to the experiences of everyday lives, in which the owners of the body become their own experts. In this way the "body is shown to be more than a logical network of biological parts or statistically verifiable behaviors—it's also an extension and motivator of unpredictable human needs and wants."[25]

The emergence of tabloid aesthetics is not restricted to print journalism but affects most modern media, which are undergoing a process of transformation. There is a new kind of public arena in which a sense of community is created, allowing access in the private home to events happening far away. At the same time the public sphere is more easily able to intrude into the private world of home and family. This blurring of boundaries, argues Lumby, means that there has been a redefinition of the kinds of issues that are worthy of public discussion. As she says "our understanding of what constitutes public debate is being rethought to include issues traditionally thought of as private [such as]...making a living and living with our families and between work and leisure."[26] In tracing the increase of private issues in public discussion, Lumby is essentially arguing that the media are drawing from previously marginalized, feminine concerns and feminine aesthetics that reveal a concern with the desires of the body. In fact, the mainstream news at times reads like a soap opera. Emotional and sexual escapades are constantly being revealed.

Thus, "feminine" aesthetics are not unique to *Buffy* and *Nikita*. Indeed, the publicizing of the private realm that Lumby describes is seen not only in the rise of tabloid journalism but is also reflected in the way soap opera narratives have actually become mainstream in the last decade. Notably, shows prominently featuring women or family concerns, like *Ally McBeal*, *Now and Again*, *Sports Night*, and *Providence* focus almost exclusively on relationships. Even shows beloved by a predominantly male audience, such as professional wrestling, are based on ever-changing story lines and evolving relationships as are more mainstream, critically acclaimed shows such as *NYPD Blue* and *ER*. *Ally McBeal* and series like *Dawson's Creek*, *90210*, *Roswell*, *Charmed*, *Felicity*, *Melrose Place*, and *Popular* draw even more obviously from soaps, suggesting that there is an audience craving for the sense of intimacy that is created by knowing the characters and being able to follow evolving storylines.[27]

However, there are some obvious differences between shows like *NYPD Blue* and *Buffy*. Unlike serialized action texts like *NYPD Blue*, in which most screen-time is spent on solving crimes and which present the emotional dramas of their characters in a manner meant to be read as "realistic," female-centered texts depict a heightened emotional realm in which communication and relationships are of the highest import. Through particular textual strategies, such as visual style and dialogue, female-centered texts move from a "realistic" manner of presentation to one that might be considered melodramatic, clearly operating in a fantasy realm in which situations and emotions are *excessive*. Certainly, the fantastic nature of the texts is one way in which the female hero is contained, yet this form of presentation also opens up space, as in fairy tales, to explore what always lies beneath an adventure—the psychological and emotional drama and the journey of identity.

Within female-centered texts, the melodramatic experiences of the lived body are inextricably linked within the narrative and imagery to themes of bodily danger. For example, each character in *Buffy* yearns for closeness with others at the same time their desires almost destroy them: Bodies are in constant danger from death, possession, penetration from stakes and fangs. This physical danger is linked to emotional danger. Friends and lovers may save your life, but when they are possessed by demons they may suddenly consume you, metaphorically and literally. As evidenced by Nikita, Buffy, and Aeon's relationships with men who help them as often as they betray them, personal connection is both necessary and desirable as well as dangerous. Moreover, when trusted friends, whom you thought you knew, are revealed to be enemies or different from what you assumed, you become implicated in the deceit. Ultimately then, in these shows the protagonist's relationship to the external, and to others, is restructured. Unlike the male hero, the female hero cannot act like a god-like figure who alters the world but plays on a more level playing field and is implicated in the danger of her own desires.

Certainly, it is not impossible to conceive of a male hero, as in film noir, who is less omnipotent than threatened. Still, there is a sense that the stakes are much higher for the female hero than the typical hero simply because emotions *are* foregrounded. In our society, women are inherently more touchable, more physically and emotionally approachable, by both women and men. Just as in the social world, women's personal space is encroached by others, the textual and narrative strategies embraced by these texts encourage viewers to move in more closely to *Nikita* and *Buffy*. Additionally, it is certainly more acceptable for females to present a range of emotions such as fear, sorrow, and pain. In this way, the femaleness of the hero allows an intensity of emotions to be expressed, spoken, and dwelt over. Importantly, these issues are not hidden or repressed as they might be with the hero, but they are externalized, brought out into the public realm, both for view and for debate.

Thus, we see in both *Nikita* and *Buffy* what I would describe almost as a fetishization of private bodily experiences, which are also apparent in other cultural forms like soaps and tabloid journalism. Tabloids promise their readers access to private truths.[28] Moreover, Lumby argues that tabloid journalism emphasizes "performance, emotion, and spectacle."[29] Similarly their narratives not only revolve around emotions but the *spectacle* of emotion, desire, and fear. Even in *Aeon Flux*, the visual imagery is concerned with showing the internal action, the subtext, represented by the movement of the virus through the bloodstream. In other words, these three series stress the primacy of the sensual moment, private internal workings, and the experience of the body located in a particular space at a particular time.

Undoubtedly, bodily sensation in the face of death has always been the focus of action texts. Westerns are filled with descriptions of the body's sensation in the environment. Heroes suffer physically against bleak, harsh landscapes. The pleasure they find in their bodily existence is thus masochistic, if anything. In contrast, while female heroes suffer in dangerous spaces, their suffering is emotional rather than physical, and physical relations, while dangerous, are also pleasurable. Obviously, this replicates the traditional association with the female to the body and the male to the mind. However, the meaning of death and the goal of heroic self-transcendence change with these different conceptions of the body. While the hero achieves transcendence through the domination and the othering of his body, the meaning of transcendence changes for the embodied female hero. If your identity is conceived of as necessarily connected to your body, how do you imagine a world beyond your body?

Within most action texts this question is suppressed, not only by repressing the fact of the hero's death but also by the way the body is conceived of as a tool. The male hero is first and foremost, himself. He acts upon his body to control it. Once the hero achieves his transcendence, the tool, like a hammer, can metaphorically, and sometimes even literally, be thrown away. But for the female hero, whose bodily sensations and relationships are presented as *so* crucial to the formation of her identity, such a split is harder to imagine.

For example, Nikita and Buffy must learn to live their lives in the face of death. Just as Nikita is an agent until she dies, Buffy is the Slayer until she dies. There is no retirement. Yet despite the potential for and prominence of death in all three series, death, at least in *Aeon Flux* and *Buffy*, is actually denied, not only by the survival of the hero, but through the presentation of death as somehow beside the point. Even when death happens, the body remains.

For instance, Buffy exists in a space between life and death in which the distinction between the two is often blurred. Buffy, in fact, dies twice only to be brought back to face the same life-based problems. A certain fluidity between life and death is suggested by the frequency with which characters move back and forth between death and life as do the vampires, zombies, Buffy, and Angel. Aeon Flux dies but is reborn in almost every episode. This is apparent in *Xena* as well with various characters moving back and forth between life and death on several occasions. Moreover, the undead or reborn bodies retain the memories, identity, and desires of the lived bodies. This concept certainly suggests some kind of openness to metaphysics, that there is something outside of life, in which there will be a continuance of identity through the body.[30]

In addition there is a very reductive quality to all three series which denies time and thus the termination of life. The characters in *Buffy* are all very much focused in the *now*, not the future. The past is useful in that it can reveal clues about how to figure out the present. However, the past is incorporated and seen as equal to the present. For example, the ancient vampires such as Angel, Spike, and the Master, all live in the now. Despite jokes about Angel "robbing the cradle," Buffy and Angel are equal in the present moment, the moments they share are sensual, emotionally intense. The anachronisms of *Xena* likewise compress time, and the use of camp in both shows continually draws upon the past to create common experience. The past "meaning" of camp references are dissolved; camp reframes and redefines their meaning based on the present moment. The past is only important in that it created *us*.

Similarly, *Aeon Flux* exists in a world in which past, present, and future blur together. It is a world of never-ending change, a world in which any sense of linearity or causality in actions is disavowed. This works to reduce all moments to the present. Likewise, *Nikita* features hyperattention to the present, sensual moment, expressed by the use of music, slowed-down filming, and narrative concerns about emotional metanarratives rather than action narratives. Plot is less important than affect; sensuality, experience, and *nowness* are highlighted over the discovery of "truth" or any ultimate meaning.

The feeling evoked from the visual (and musical) attention to the sensual moment in all three series is such that the momentary, lived experience of the body achieves timelessness. Kierkegaard has said that "[t]he instant is that ambiguous moment in which time and eternity touch one another, thereby positing the temporal, where time is constantly intersecting eternity and eternity constantly permeating time."[31] In other words, the lived experience of the body, before it is reflected upon, at least, is a way the living being can connect to a sense of immortality.

Likewise, in her examination of another female-focused action text, Linda Badley has observed that *The X-Files*' treatment of the relationship between death, truth, and the body suggests "transcendence of the body through the medium of the body."[32] As in *Buffy* and *Aeon Flux*, "no body really dies." Badley observes that Foucault's writing on the clinic reframed the meaning of death. "The new mortalism produced by the clinical perspective...promised that the body could be manipulated, its processes reversed or suspended."[33] Further, the characters of Scully and Mulder represent conflicting viewpoints, mind and body, that increasingly meet in the liminal zone of the series. "They connect, like two halves of the same person...in the text of *The X-Files*...[a] fantastic space between the empirical and metaphysical. Together, and obsessed like Victor Frankenstein...by the urgency of their work, they convert the physical to the metaphysical."[34]

Of course, the after-death survival of the individual has been a major theme in American religion since the mid-to-late nineteenth century. That there is a widespread desire for eternal life in which the individual continues to exist as a discreet entity is certainly suggested by the popularity of psychics and stories of near-death experiences which posit an afterlife much like this one. However, the sense of the immortality of the individual as being *necessarily* connected to the body seems to me to be particular to the contemporary age. Mike Featherstone has described how contemporary consumer culture has resulted in a new conception of self based on the signs of the body.[35] Likewise, in her study of girls' diaries, Joan Brumberg has observed that, whereas in the past girls gained a sense of self through good works and a sense of morality existing outside the body, contemporary girls base their identities on their bodies.[36] At the same time we have an aging population and have entered into a new millennium, it seems hardly coincidental that larger cultural discourses have begun to privilege the experiences of the body.

David Morris similarly argues that the postmodern era has a radically different view of the body, as seen in how metaphors of illness have changed since the end of the nineteenth century. While previous eras

> viewed the body as a machine-like carapace of a well-fortified castle, fending off external threats from hostile bacteria and viruses, postmodern culture shows us something quite different with its vision of selves and bodies newly vulnerable to the workings of our own immune systems. It is questioning the nature of the self that falls ill, the self that is now increasingly fragile and incohesive.[37]

As Morris admits, cultural fears over disease have made regular *fin-de-siècle* appearances. Although the cultural discourse at the end of the nineteenth century conceived of illness as something without (i.e., linked to immigrants), the modern threat, both physically and culturally, is the apparent lack of control on the part of the individual or the culture over the internal workings of the body. This lack of control threatens the self since the self, the body, is implicated in its own destruction. Culturally, of course, there is a similar idea that the advances of our modern age may lead to the dissolution of society. One thinks not only of weapons of destruction like the bomb and germ warfare but the concern over the increasing numbers of resistant strains of bacteria and viruses that have developed in response to modern medications.

Interestingly, Morris sees the postmodern body as being less representative of a denial of death than of the *processes* of dying.[38] Death, he observes, is at least "photogenic," providing a spectacle for the viewer, while chronic illness and dying, decay, and the process of bodily dissolution are more fearful. Perhaps this is due to the fact that the moment of illness and decay is when we consciously become aware of our minds as being separate from the body.[39] The mind must watch helplessly as the body dies. Thus the potential death of the female hero may be less fearful than her troubled existence.

Similarly, Susan Bordo argues that the workings of the body are threatening partly because they are *not* static (like death, perhaps) but are part of a process. This process continually threatens to erupt into bulges of fat, requires constant maintenance to defend against the threat of decay (i.e., fat, wrinkles) that originates from within. Likewise, the body functions as a sign of its owner's individuality and will power. It is commonly believed that you get fat when you "lose self-control," the control of your mind over your body. Obviously, as in the case of anorexics, such mental control over the body is a way to reestablish a sense of order and control over a body gone mad. In addition, by denying its own materiality and historicity the "plastic" postmodern body Bordo describes can almost be seen as immortal as the "soul" that shaped it.

Certainly the growing fascination with plastic surgery, the increasing cultural obsession with youth, and the marketing of products like "Everlasting Youth" emerge from a culture fearful of aging and of death which desires an ever-youthful, ever-sensing, and feeling body. Everywhere around us we have images not only of youth and beauty but sex and *fun*. Particularly, there are images of the body as immortal, young and strong. However, this is not *just* a case of the beautiful young body as sign. Instead, the interiority of the body is also a central focus.

For example, shows and novels about immortal protagonists are everywhere as in *Forever Night, Angel, The Kindred*, and *Blade. Highlander* and *The Raven* feature a cast of immortal characters obsessed with their human relationships. The male and female heroes of *Roswell* are powerful aliens who have come back from the dead through being cloned. Popular characters in the *Star Trek* armada, like Spock, Data, Odo, and Seven-of-Nine, are all extrahuman, long lived, enhanced, and, except for Spock, they all strive to be "human" and understand the emotions and sensations of human life.

Perhaps the most obvious example of this tendency can be seen in the current popularity of vampires. Vampires are, like superheroes, forever young and strong. For years, they have represented sexuality, danger, exoticism, but it is only in recent years that the vampire has become one of *us*. In her examination of Anne Rice's famous vampire series, Linda Badley sees them as having "obvious appeals to a culture based in the body, the inversions of the Christian sacraments merely remind us of the values we really live by. They are the body's: predation and symbiosis, blood or life consumed and given, biological immunity as a state of grace."[40]

What is most relevant for discussion here is the route vampires travel to achieve their "biological immunity." In Rice's world, vampires, needing blood to live, go to the heart of life to find sustenance; like babies they drink from a punctured, feminized body. Like mothers, they give birth to other vampires by allowing their offspring to drink from their bodies. Most importantly, their immortality paradoxically makes them completely dependent upon the body, constantly craving blood, their waking hours fiercely regulated. The vampires' powers, their strengths, are thus inextricably tied to their frailty; their internal strength created by the influx of tainted blood. Their permeability creates their impermeability

Further, it is interesting that Rice fills her novels with descriptions of physical sensations, impressions, and emotions about particular moments, far more than any action sequence. Likewise, Badley observes that Rice's hero Lestat, exists in "a world in which distinctions between history and reality have collapsed."[41] This collapse is also apparent in *Aeon Flux*, in which causality and motive are completely missing, and in *Buffy* and *Nikita*, in which *nowness*, the experiences of the body and of consciousness in the lived moment, are paramount. As Badley argues, within her works Rice "restored and validated Eros, polytheism, substance, the primary knowledge of the sense, and an aesthetic based in desire and the body."[42]

Comparable themes occur in the *Star Trek* movie, *Insurrection*, where the crew encounters a planet full of immortals who have the power to slow down time and freeze individual moments of happiness. Television ads frequently rely on the use of fragmented, timeless, MTV imagery to sell their products. *Alien*'s Ripley even comes back from death as a clone, though her DNA has been blended with that of the alien. The cloning, which is presented in the film as an assault, a raping of Ripley's bodily integrity, echoes the paradoxical existence of the vampires, since the infiltration of alien DNA gives her the abilities and heightened alien sense she needs to survive. The female hero of *Dark Angel* likewise is the result of a government experiment, her human DNA having been mixed with that of a cat. Her superhuman strength is both the source of her survival and the cause for debilitating seizures and her continued persecution by government forces.

Outside of science fiction and fantasy, the air waves are full of talk shows like *Jerry Springer, Sally Jessy, Dr. Laura*, and *Dr. Joy* in which individuals confess their transgressive bodily behavior. Exposing themselves, they allow viewers into the intimate, vulnerable places of their psyches, but at the same time they elevate their mundane lives into the eternal existence conferred by fame. Bodies become the route to immortality. However, it is important to note that theoretically these are *feminized* bodies; they are opened, threatened, penetrated, and still they survive—in fact, they *thrive*.

Badley notes the obvious appeal for Rice's series among gay communities as well as women is based on the "feminized" bodies of the vampires. Rice's vampires are androgynous, the male vampires bonding with each other in relationships that are intimate and nurturing.[43] Obviously the homoerotic qualities offer appeal, while androgyny promises power without the limitations of gender roles. Similarly, women and gay men have "feminized bodies" vulnerable to sexual penetration, and as Badley observes, the threat of AIDS. Vampire tales and female heroes reveal dreams of bodily weakness mutating into strength.

Obviously then, the use of the body to achieve transcendence need not be limited to the feminine. But in a culture in which the body is most often associated with the feminine and which is correspondingly particularly threatened by death, it is not surprising that visual culture has begun to feature the survival and transcendence of the female hero. The embodied quality of the feminine evokes both the potent sensuality and the threatened fragility of lived experience. In fetishizing the impermanent, the now is transformed into the forever. By doing so, the body as *being* becomes static, frozen in time, a spectacle, shored up against the process of decay.

Thus, these series demonstrate signs of a cultural restructuring of conceptions of the body. Perhaps stemming from an age-old hysteria around the weakness of the body, the body is simultaneously revealed as a source of strength and the path to immortality—not by leaving the body behind, but by existing within it. And while this leads on the one hand to images of immortality and an individuality that will never die, on the other hand, it also invests the body with relevance—a sense that its workings and processes, even its very vulnerability, are precious.

EPILOGUE

Female Heroes and Postfeminism

This study began with the question of what the new representations of the female body have to say about women's claiming of public space and consequently women's power in culture. What makes these texts interesting is that they are so contradictory: It is impossible to reduce the "meaning" of any of these texts to one position. They feature women who effortlessly take up a great deal of space: No one is a physical match for Buffy; Nikita eventually takes over Section; and Aeon regularly defies the laws of physics. Yet in relation to their environments, their bodily boundaries are open and fluid, a representation that can be read as both a strength and a weakness. The politics of the series are likewise frequently contradictory, critiquing the notions of individuality and dominance seen in many male-centered texts at the same time they often reinscribe them. *Buffy*, for instance, features an ethics of care; at the same time it promotes a worldview built around the necessity of fighting and domination.

Ultimately, it depends on one's perspective of whether these series are visions of "feminist progress" or not. As Patricia Pender writes of *Buffy*, taking an either/or view of the "feminism" of a text is less productive than viewing texts as sites in which cultural negotiations over the meaning of terms like feminism and female power take place.[1] This final section considers some of the ways in which these series are simultaneously transgressive and regressive in their presentation of the female hero. Although recognizing the various ways these texts reconfigure the traditional action narrative, it is important to critique those elements of the series which remain politically regressive.

Certainly, all three series offer far more positive role models for girls than have previously existed. Although sexually appealing, Buffy, Aeon, or Nikita cannot be read as *just* sex symbols. In addition, each of these series offers significant ways for the female viewer to insert herself within the text. As Green has observed in regards to *Xena*, though the intended audience and camera's gaze might be male, the space of the text presents a female gaze.

Male viewers, therefore, can't enjoy the female hero's humiliation, though they can enjoy the hero's struggle as representative of the individual underdog.[2] The presence of the female gaze in the visual space that is apparent in *Xena*, *Buffy*, *Nikita*, *The X-Files*, and *The Net* seems representative of feminist inroads into culture. Moreover, the potentially positive psychic effects on their viewers cannot be denied.

And while all television female heroes remain white, pretty, thin, mainly straight, and not particularly muscular, they still offer hope of a different way of seeing. As Green has also noted, the different depictions of the relations between individual and community in female-centered texts can create a spirit of egalitarianism that is different in tone from most male-centered texts.[3] Further, the focus on the individual moment and sensation in the texts under examination here is different than traditional male-centered texts in that it places feeling and impression (erotic knowledge) rather than narrative action (climactic knowledge) at the center of importance.

However, as previously noted, Green concludes that ideologically, at least, most mainstream female heroes replicate traditional concerns with individualistic success and do not alter the basic American epistemology in any constructive "feminist" way. I don't deny this. Heroism is dependent on overcoming obstacles. In this way, social pressures are naturalized and normalized within action texts. Yet it must also be said that traditional heroic action/adventures are *about* the discovery of identity; notions of individualism are somewhat inseparable from the genre. When traditional modes of representing action narratives are broken with, as they are in *Aeon Flux*, there is a corresponding relativism to her character that denies transcendence. Aeon is pathology itself. While Aeon may offer an image of extreme female power to her viewers, she is not really a *hero* in the sense of serving as a role model or figure to emulate.

In addition, the meaning of individuality has always been and is still different for women than for men. Individuality is, after all, about control, both of the self and of others. While traditional heroic texts foreground the establishment of control over others (the ultimate goal of bodily control), female-centered texts primarily articulate concerns/tensions with autonomy, issues over gaining and maintaining control over the self. In other words, female control is about *establishing* the independence and self-determination with which heroes begin. Nikita desires her freedom from Section and wants to find a way to be happy in her life. Buffy similarly hates her life as Slayer and then begins to seek mastery over the terms that control her life. The distinction holds true for *Xena* as well.[4] These women are fighting against being overwhelmed, losing their individuality, having their bodies taken from their personal control.[5]

Clearly the images at play here might have a certain appeal for many women. They function as feminist critiques in that heroes like Buffy are fighting "male dominant authority structures."[6] Likewise, the endangered state of the female hero is a way of criticizing male violence as well as justifying the use of female violence. Additionally, the hero's fight for bodily control corresponds to the genre's usual articulation of the tension between community and individuality as the threat of being swallowed up by the external world, the demands of others, resonates with public discussions over contemporary working women. According to mainstream media and shows like *Ally McBeal*, for example, female workers are having a horrible time: they are overworked both at home and at work and exhausted by the many roles they have to occupy. Similarly, Buffy and Nikita have jobs that they often hate, pull them in different directions, and wear them down. They, and heroes like Scully and *Profiler's* Sam, try to balance their private lives with the demands of their jobs. Obviously the stress they feel may resonate with female viewers.

However, neither series provides a truly feminist response to the hero's plight. Just as much public discourse centers on the need for women to either shut up or make a "choice" they often don't have (i.e., to work *or* be wives and mothers), Buffy and Nikita are repeatedly forced to silence their protests and discover a way to survive in a hostile environment, to change themselves for a world which will not change *for* them. The hero's adversarial relationship to her external world and the fetishization of private bodily experience at times convey an ideology associated with the politics of "postfeminism."

Postfeminists like Katie Roiphe have distanced themselves from the concern that feminists of earlier years had with documenting women's oppression at all levels of society. They argue against what they perceive as an outdated insistence by other feminists on the 'victimization' of women, believing that it is the responsibility of the individual to empower herself and stop blaming cultural factors for women's oppression. Their call is thus for individual, not social, change. Not surprisingly, postfeminists, women who are engaged with self-growth rather than social change, have been presented by the mainstream media as offering an "acceptable" version of feminism. The media popularity of the postfeminist position is due to the fact that it is seductive. It appeals to deeply imbedded American ideologies of individuality. It reassures people who want to believe that the world is orderly and can be controlled; if you fail, if you suffer, it's *your* fault. This is a brand of feminism sure to be easily digested by television, a medium that prefers middle-of-the-line political positions and projects happy, middle-class whiteness as the norm. It therefore benefits all those who want to maintain the status quo as well as those who are just plain scared.

Writers like Roiphe have been critiqued, mocked, and condemned by a host of other, less famous feminists, whose ideologies remain less friendly to the white, male-dominated media world. Feminists such as bell hooks note that the idea of individual power tends to be promoted by those who are privileged, wealthy, white, educated women possessing advantages not held by the majority of women in the world. These writers focus on the continued need for wide-based political movements among groups of women and men to bring about positive social change.

One could argue that the "power" feminism of "postfeminists" is better than no feminism; at least there is an assumption that *some* women are equal to men. But I'm not so sure the advantages outweigh the disadvantages. The danger of this position is, as already noted, that it erases real-life differences (poor women, for example, do *not* have the same access to the monetary and social resources held by middle-class women), often works to blame the victim, and ultimately suggests a cult of individualism that lacks the possibility for effective social change.

Buffy offers a particularly potent example of one such danger. Although the series' emphasis on an ethics of care is transgressive on many levels, its representation of sex and desire throughout six seasons is deeply repressive. The series repeatedly depicts the risks that threaten teen-aged girls yet promotes the idea that strong women are responsible for controlling *themselves*. This self-control, in turn, has the effect of containing the outside dangers that threaten them. It's Buffy's responsibility to avoid having sex with Angel or Spike lest they do evil. Willow must be willing to anesthetize Oz, lest he go on a murderous rampage; she must control her magical abilities and her desire for power lest she be taken over by dark magic. Girls who fail to control their desires, like Faith, or fail to control their lovers, like Debbie, hurt not only themselves but others. These women owe it to society to protect themselves.

Part of what disturbs me about *Buffy*'s emphasis on female self-control is that it is repeatedly echoed in other arenas. For example, in the popular radio show, television series, and self-help book, *Loveline*, women are repeatedly told that they are the "keepers of intimacy" in a relationship. The language of the (white male) hosts often stresses the need for women to protect themselves by taking responsibility for their own actions. This in itself is practical advice and is undoubtedly intended to encourage women to have self-respect. However, the result of all the lecturing and cajoling for women to stand up for themselves is that men, who are the perpetrators of abuse, are not held responsible for their actions. It becomes just another way of blaming the victim and excusing the victimizer, while making abusive behavior seem the norm.[7]

While I don't think this is the conscious intent of the *Loveline* hosts, it is certainly the frequent effect of their rhetoric. Such discussions are certainly not unique. They emerge, for example, in much current discourse about campus date rape—which cautions women to avoid placing themselves in unsafe situations—as if female self-control will solve the problem, when the real problem is that men feel entitled to *take*. Contrary to what some think, promoting personal power and individual responsibility as the solution to all social ills is hardly revolutionary or new. Instead, such arguments merely replicate previous ideologies in which the fittest individuals triumph and in which women may protect themselves against the threat of rape by blaming other women for not being "empowered" enough to stand up for themselves.

It is interesting that a text like *Aeon Flux* appeared in the nineties in the era of Generation X and postmodernism, a time in which distinctions between the real and imaged, reality and illusion, have been effaced.[8] The flight from social responsibility demonstrated by "power feminists" certainly seems relevant to much of our contemporary postmodern society. As seen through media images, particularly ads marketed towards Gen X'ers, the current age is a world that is so confusing and paradoxical that it is not worth thinking about and you might as well relax. "Just do it," says Nike, or in the words of a Mountain Dew commercial, "Do the Dew." This is a time where action comes before thought since all thought is useless and confusing and only leads to more paradoxes. Action and life have value, not because of any intrinsic quality, effect, use, or purpose (thus there is no real need for politics) but just because they *are*.

The focusing on instant gratification without any consideration of consequences has the same effect as nihilism. The denial of material reality—the fact that our existence is prescribed by certain factors, such as social conditions, and limited by our mortality—also suggests a denial of weakness, a denial of death.[9] Strength is found in consuming, pleasuring the self, as opposed to social reflection or self-reflection. *Aeon Flux* promotes a similar disdain for compassion. It is a postmodern text in that it layers images upon images in such a way as to deny any one clear-cut meaning. However, this lack of fixity can be problematic; one critique of postmodernism is that, like postfeminism, it erases differences between people and dissolves social responsibility.

By focusing on individual empowerment or individual *fun*, the needs of others are dismissed. In her article "And Now, Babe Feminism," Anna Quindlen quotes a postfeminist "babe," Lois Maffeo, as saying "A lot of us just want to go spray paint and make out with our boy friends and not worry about oppression."[10] In this guise, "power" feminism is quite *irresponsible*. These arguments present women as somehow living on a level playing field that does not, quite simply, exist for the vast majority of women and other oppressed people. Quindlen responds to Maffeo's desire to spray paint, "Cool—that'll make it a lot easier when you get a straight job and get paid a whole lot less than the guy you work next to."[11] Spray painting and making out don't do much in terms of transforming society. The rhetoric of fun and of postfeminism is thus just a new term to promote conservative values and behaviors which play into consumption.

However, while there are undoubtedly problems with the postfeminist idea that women need to "step up" to the male paradigm in order to have equality, it is undeniably important that Buffy dramatizes the contradictory reality of the way women experience their bodies; Nikita is at least resisting her oppression, and Aeon Flux escapes containment by the rational male gaze. While Nikita still swoons over the ambiguous behavior of an often-cruel Michael, in *Buffy* bad boy behavior is clearly not acceptable. Though it is "natural" for guys to be beasts, if they don't see the problem with their beastly behavior, Buffy will slay them. It is at least *implied* that men should work on their own behavior and save women the trouble.

The three series here thus mediate a variety of different strands of thought within our contemporary culture. Tales of action and female bodies are being reconfigured, opening up spaces for female power and physicality, female individuality and heroism, and also functioning as expressions of feminine discontent and feminist critique. Yet the female hero remains a contradictory figure. The need for collective social action, though apparent in female-centered texts, is subsumed in overarching themes of individualism. And while the white, female bodies of these heroes are comparatively empowered, they are explicitly presented as embattled and under threat of penetration.

However, the hero's vulnerable body may be due less to how people perceive female heroes, than to changing cultural attitudes towards the body. In other words, the female hero's embattled body may merely signal a cultural view in which the body is seen as vulnerable, exposed, in which boundaries between public and private are broken and which transforms the myths of heroism and the meaning of transcendence. The focus in these series is in not only learning how to survive in the here and now but how to appreciate the experience. Yet at the same time it seems that it is harder and harder for us to imagine that there will not always be a *now*—that we will be gone.

NOTES

Introduction

1. M. Ann Hall, *Feminism and Sporting Bodies: Essays on Theory and Practice* (Champaign, IL: Human Kinetics, 1996): 50.
2. Maggie Hadleigh-West, *War Zone* (Northampton, MA: Media Education Foundation, 1998), videocassette.
3. John Berger, *Ways of Seeing* (London: Penguin, 1972).
4. Iris Marion Young, *Throwing like a Girl and Other Essays in Feminist Philosophy and Social Theory* (Bloomington: Indiana University Press, 1990): 141–160.
5. Carole Spitzack, *Confessing Excess: Women and the Politics of Body Reduction* (Albany: State University of New York Press, 1990); and Susan Bordo, *Unbearable Weight: Feminism, Western Culture and the Body* (Berkeley: University of California Press, 1993).
6. Kathleen Rowe, *The Unruly Woman: Gender and the Genres of Laughter* (Austin: University of Texas Press, 1995).
7. Laura Mulvey, "Visual Pleasure and Narrative Cinema," in *Contemporary Film Theory,* ed. Antony Easthope (New York: Longman, 1993): 111–125.
8. Bonnie Dow, *Prime-Time Feminism: Television, Media Culture, and the Women's Movement since 1970* (Philadelphia: University of Pennsylvania Press, 1996).
9. Philip Green, *Cracks in the Pedestal: Ideology and Gender in Hollywood* (Amherst: University of Massachusetts Press, 1998): 11.
10. Green, *Cracks in the Pedestal*, 188.
11. Bordo, *Unbearable Weight*; and Susan Willis, *A Primer for Daily Life* (New York: Routledge, 1991): 62–85.
12. Henry Jenkins, *Textual Poachers: Television Fans and Participatory Culture* (New York: Routledge, 1992).
13. Anne Balsamo, "Forms of Technological Embodiment: Reading the Body in Contemporary Culture," in *Cyberspace, Cyberbodies, Cyberpunk: Cultures of Technological Embodiment*, eds. Mike Featherstone and Roger Burrows (London: Sage, 1995): 217–218.
14. Norman Fairclough, *Language and Power* (New York: Longman, 1989).
15. Julie D'Acci, *Defining Women: The Case of Cagney and Lacey* (Chapel Hill: University of North Carolina Press, 1994).
16. Elizabeth G. Traube, *Dreaming Identities: Class, Gender and Generation in 1980's Hollywood Movies* (Boulder: Westview Press, 1992): 14; and Lynn Segal, *Slow Motion: Changing Masculinities, Changing Men* (New Brunswick, NJ: Rutgers University Press, 1990): 92.

Chapter One: Feminist Media Theory

1. Mulvey, "Visual Pleasure and Narrative Cinema," 111–125.
2. Ibid.
3. Tania Modleski, *Loving with a Vengeance: Mass-Produced Fantasies for Women* (New York: Routledge, 1982).
4. Linda Williams, "'Something Else Besides a Mother': *Stella Dallas* and the Maternal Melodrama," in *Issues in Feminist Film Criticism*, ed. Patricia Erens (Bloomington: Indiana University Press, 1990); and "Feminist Film Theory: *Mildred Pierce* and the Second World War," in *Female Spectators: Looking at Film and Television*, ed. E. Deidre Pribram (London: Verso, 1988).
5. Traube, *Dreaming Identities*, 11.
6. Janice Radway, *Reading the Romance: Women, Patriarchy, and Popular Literature* (Chapel Hill: University of North Carolina, 1984).
7. Molly Haskell, *From Reverence to Rape;* and Majorie Rosen, *Popcorn Venus: Women, Movies, and the American Dream* (New York: Coward, McCann & Geohage, 1973).
8. Ellen Seiter et al., "Introduction," in *Remote Control: Television, Audiences and Cultural Power*, ed. Ellen Seiter, Hans Borchers, Gabriele Kreutzer and Eva-Maria Warth (London: Routledge, 1989).
9. Tania Modleski, "Introduction," in *Studies in Entertainment*, ed. Tania Modleski (Bloomington: Indiana University Press, 1986).
10. Patricia Mellencamp, *High Anxiety: Catastrophe, Scandal, Age and Comedy* (Bloomington: University of Indiana Press, 1991): 273.
11. Bonnie Dow, *Prime-Time Feminism*, xx.
12. Ibid., 6.
13. Ibid.; and Green, *Cracks in the Pedestal*, 6.
14. Cited in Julie D'Acci, *Defining Women*, 3.
15. Dow, *Prime-Time Feminism*, 5 .
16. Vickie Rutledge Shields with Dawn Heinecken, *Measuring Up: How Advertising Affects Self-Image* (Philadelphia: University of Pennsylvania Press, 2002).
17. Dow, *Prime-Time Feminism*, 11.
18. Ibid., 12.
19. Jim Collins, *Uncommon Cultures: Popular Culture and Post-Modernism* (New York: Routledge, 1989): 42.
20. Dow, *Prime-Time Feminism*, 7–8.
21. Traube, *Dreaming Identities*, 5.
22. Ibid.
23. John Fiske, *Television Culture* (London: Methuen, 1987): 5–52. Cited in Dow, *Prime-Time Feminism*, 12.
24. Dow, *Prime-Time Feminism*, 12.
25. Ibid.; Celeste Condit, "The Rhetorical Limits of Polysemy," *Critical Studies in Mass Communication* 6 (1989): 103–122.
26. Dow, *Prime-Time Feminism*, 13.
27. Ibid., 19.
28. Ibid., 19.
29. Ibid., 15.
30. Ibid., 14.
31. Ibid., 4.
32. Ibid., xix.

33. Traube, *Dreaming Identities*, 11. Emphasis mine.
34. Christine Gledhill, "Developments in Feminist Film Criticism," in *Re-Vision: Essays in Feminist Film Criticism,* eds. Mary Ann Doane, Patricia Mellencamp, and Linda Williams (Frederick, MD: University Publications of America, 1984); and Traube, *Dreaming Identities*, 14.
35. Dow, *Prime-Time Feminism*, xiv.
36. John Fiske and John Hartley, "Bardic Television," in *Television: The Critical View*, ed. H. Newcomb (New York: Oxford, 1987): 602. Cited in Dow, *Prime-Time Feminism*, 8.
37. Dow, *Prime-Time Feminism*, xv.
38. Ibid., 8.
39. Ibid., xvi.
40. Williams, "Feminist Film Theory," 28.
41. D'Acci, *Defining Women,* 5.
42. Traube, *Dreaming Identities*, 14.

Chapter Two: Interpreting Heroic Action

1. Ginia Bellafante, "Feminism: It's All About Me!" *Time*, 29 June 1998, 54–62.
2. Michael A. Messner, "Sports and Male Domination: The Female Athlete as Contested Ideological Terrain," in *Philosophic Inquiry in Sport* (2nd ed.), ed. William Morgan, Klaus Meier (Champaign, IL: Human Kinetics, 1995): 278.
3. Bordo, *Unbearable Weight.*
4. Naomi Wolf, *The Beauty Myth* (New York: Anchor, 1992).
5. Kathleen Rowe, *The Unruly Woman,* 25–49.
6. Patricia Mellencamp, *High Anxiety.*
7. Rowe, *The Unruly Woman*, 52.
8. Julie D'Acci, *Defining Women*, 16.
9. Ibid., 16.
10. Green, *Cracks in the Pedestal*, 211.
11. Green, *Cracks in the Pedestal*, 188.
12. Ibid., 183.
13. Ibid., 182.
14. Ibid., 183. Emphasis mine.
15. Yvonne Tasker, *Working Girls: Gender and Sexuality in Popular Cinema* (New York: Routledge, 1998): 68–69.
16. Yvonne Tasker, "Fists of Fury: Discourses of Race and Masculinity in the Martial Arts Cinema," in *Race and the Subject of Masculinities*, eds. Harry Stecopoulos and Michael Uebel (Durham: Duke University Press, 1997): 332.
17. Green, *Cracks in the Pedestal*, 177–178.
18. Kathleen Rowe, "Roseanne: Unruly Woman as Domestic Goddess," in *Feminist Television Criticism: A Reader*, eds. Charlotte Brunsdon, Julie D'Acci and Lynn Spigel (Oxford: Clarendon Press, 1997): 77.
19. Yvonne Tasker, *Spectacular Bodies: Gender, Genre, and the Action Cinema* (New York: Routledge, 1993): 30.
20. Ibid., 3.
21. Ibid., 3.
22. Ibid., 5.
23. Ibid., 115–116.

24. Tasker, *Working Girls*, 70.
25. Ibid., 70.
26. Ibid., 69.
27. Ibid.
28. Ibid.
29. Tasker, *Working Girls*, 71.
30. Radway, *Reading the Romance*.
31. Tasker, *Working Girls*, 82.
32. Sherrie A. Inness, *Tough Girls: Women Warriors and Wonder Women in Popular Culture* (Philadelphia: University of Pennsylvania Press, 1999): 161.
33. Ibid., 160–176.
34. Ibid., 176.
35. Balsamo, "Forms of Technological Embodiment," 217–218.
36. Jane Tompkins, *West of Everything: The Inner Life of Westerns* (New York: Oxford University Press, 1992): 4.
37. Ibid.
38. Ibid.,15.
39. Ibid., 45.
40. Ibid., 31.
41. Ibid., 17–18.
42. Ibid., 56.
43. Ibid., 57.
44. Ibid., 87.
45. Ibid., 106.
46. Ibid., 109.
47. Ibid., 122.
48. Ibid., 30–38.
49. Ibid., 31.
50. Ibid., 76.
51. Ibid., 85.
52. Susan Jeffords, *Hard Bodies: Hollywood Masculinity in the Reagan Era* (New Brunswick, NJ: Rutgers University Press, 1994): 27.
53. Ibid., 55.
54. Tasker, "Fists of Fury," 315–316.
55. Ibid.
56. Jeffords, *Hard Bodies*, 49–50.
57. William Warner, "Spectacular Action; *Rambo* and the Popular Pleasures of Pain," in *Cultural Studies*, eds. Lawrence Grossberg, Cary Nelson, and Paula Treichler (New York: Routledge, 1992): 676.
58. Jeffords, *Hard Bodies*, 170.
59. Ibid., 176.
60. Tasker, *Spectacular Bodies*, 83.
61. Tasker, "Fists of Fury," 319.
62. Ibid., 315–316.
63. Ibid., 321.
64. Ibid., "Fists of Fury," 321.
65. Joseph Lenz, "Manifest Destiny: Science Fiction and the Classic Form," in *Coordinates: Placing Science Fiction and Fantasy*, eds. George E. Slusser, Eric S. Rabking, and

Robert Sholes (Carbondale: Southern Illinois University Press, 1983); and Anne McClintock, *Imperial Leather: Race, Gender and Sex in the Colonial Conquest* (New York: Routledge, 1995).

66. Barbara Creed, "Alien and the Monstrous-Feminine," in *Alien Zone: Cultural Theory and Contemporary Science Fiction Cinema*, ed. Annette Kuhn (London: Verso, 1990): 215.
67. Susan A. George, "Not Exactly 'of Woman Born': Procreation and Creation in Recent Science Fiction Films," *Journal of Popular Film and Television* 28 (2001): 176–184.
68. Peppers, "'I've Got You Under My Skin,'" p. 166

Chapter Three: *La Femme Nikita*, Romance, and the Plastic Body

1. Julie D'Acci, *Defining Women: The Case of Cagney and Lacey* (Chapel Hill: University of North Carolina Press, 1994).
2. Jonathon Bernstein, "Cherchez la Femme," Gear, September–October 1998, 72.
3. Ibid., 74, 78.
4. Ibid., 72.
5. "Drop Dead Looks" [on-line]. 16 February, 1999. Available from World Wide Web: (http://www.nikita.com)
6. This sense of victimization is addressed by several reviewers: see Matt Roush, "The Roush Review," TV Guide, 28 March–3 April 1998, 9; and Tom Carson, "La Femme Dilbert," *The Village Voice*, 13 January 1998, 77.
7. Kirsten Marthe Lentz, "The Popular Pleasures of Female Revenge (or Rage Bursting in a Blaze of Gunfire)," *Cultural Studies* 7, no. 3 (1993): 379.
8. Michel Foucault, *Discipline and Punish: The Birth of the Prison* (New York: Random House, 1995): 200–201.
9. Bordo, *Unbearable Weight*, 245–246.
10. Sandy Flitterman-Lewis, "Psychoanalysis, Film, and Television," in *Channels of Discourse, Reassembled: Television and Contemporary Criticism* (2nd ed), ed. Robert C. Allen (Chapel Hill: University of North Carolina Press, 1992): 232.
11. Ibid., 228.
12. Ibid.
13. Ibid.
14. Ibid., 226.
15. Available on-line at http://www.nikita.com.
16. John Cawelti and Bruce A. Rosenberg, *The Spy Story* (Chicago: The University of Chicago Press, 1987): 13–14.
17. Maria Pini, "Women and the Early British Rave Scene," in *Back to Reality? Social Experience and Cultural Studies*, ed. Angela McRobbie (Manchester: Manchester University Press, 1997): 159.
18. Susan McClary, Feminine *Endings: Music, Gender and Sexuality* (Minneapolis: University of Minnesota Press, 1991): 155, 165.
19. Ibid., 165.
20. Tara O'Shea, "The Cat," 6 April 1999. [on-line]. Available from World Wide Web: (http: //www.geocities.com/-teamhair/cat.html)
21. Ibid.
22. Radway, *Reading the Romance* (Chapel Hill: University of North Carolina, 1984): 134.

Chapter Four: *Aeon Flux* and the Grotesque Body

1. Notebook, *Time Magazine*, 22 July 1996, 19.
2. Sony Playstation, December 1996; PC Version available from Cryo/Viacom New Media, January 1997.
3. Japhet Asher and Peter Gaffney, *The Herodotus File* (New York: MTV Books/Pocket Books, 1995).
4. *Aeon Flux: Mission Infinite* (New York: MTV Network/Sony Music Entertainment, Inc./ Viacom International, 1997) videocassette; *Aeon Flux: Operative Terminus* (New York: MTV Network/ Sony Music Entertainment, Inc./ Viacom International, 1998), videocassette.
5. News Brief, *SciFi Wire: A News Service of the Sci Fi Channel*, 28 June 1999. [on-line] Available from World Wide Web: (http://www.scifi.com).
6. Colson Whitehead, "That Girl," *The Village Voice*, 22 August 1995, 43; and Michael Atkinson, "Inside Aeon," *Film Comment*, 1 January 1996, 20.
7. Sut Jhally, *Dreamworlds II: Desire, Sex, Power in Music Videos* (Northampton, MA: Media Education Foundation: 1995), videocassette.
8. *neXus siX magaZine*, 1995/1996. [on-line] 15 May 1999. Available from World Wide Web: (http:// www.hallucinet.com/n6/aeon.html).
9. Peter Chung, in a message posted to *AOL's Aeon Flux Message Board*, 24 September 1995 [on-line]. 21 August 2002. Available from World Wide Web: (http://members.aol.com/lalinton/from.htm).
10. Peter Chung, in an interview with Michelle Klein Häss, 26 January 1997. [on-line] 26 January 1997. Available from World Wide Web: (http://www.stealth.net/~ch/af/mkhpc.htm).
11. *neXus siX magaZine*, 1995/1996.
12. *neXus siX magaZine*, 1995/1996.
13. Japhet Asher and Peter Gaffney, *The Herodotus File* (New York: Pocket Books, 1995).
14. Michel Foucault, *Discipline and Punish: The Birth of the Prison*. Trans. Alan Sheridan (New York: Random House, 1995).
15. Bordo, *Unbearable Weight*.
16. Rowe, *The Unruly Woman: Gender and the Genres of Laughter* (University of Texas Press: Austin, 1995): 33.
17. Ibid., 32.
18. Ibid., 32–33.

Chapter Five: *Buffy the Vampire Slayer* and the Body in Relation

1. "50 Greatest Shows of All Time," *TV Guide Magazine*, [on-line] 6 August 2002. Available from World Wide Web: (http://www.tvguide.com/magazine/issues/020429/poll.asp).
2. Max Alexander, Robin Honig and Ethan Alter, "50 Greatest Shows of All Time," *TV Guide*, 4–10 May 2002, 18–55; and "New AFI Awards Show Airs Saturday," *CNN.com*, 4 January 2002. [on-line] 15 August 2002. Available from World Wide Web: (http://www.cnn.com/2002/SHOWBIZ/News/01/04awards.afi.advance).
3. Ien Ang, *Watching Dallas: Soap Opera and the Melodramatic Imagination* (London: Methuen, 1985): 83.
4. "The O'Reilly Factor," Fox. 26 April 1999.

5. Joss Whedon, "Interview, Welcome to the Hellmouth [video]" cited by Kathleen McDonnell, "Chaos at the Mouth of Hell," *Gothic Studies* 2, no. 1 (2000).
6. Ad appears on the back cover of *Sci-Fi Teen*, November 1998.
7. Materials available from Power Star Collectibles, Audubon, NJ, and Federation Comics, Reading, MA.
8. Stephanie Zacharek, "Willow, Destroyer of Worlds, " *Salon.com* [on-line] 30 May 2002. Available from World Wide Web: (http://www.salon.com/ent/letters/2002/05/30/buffy/index.html).
9. Diane DeKelb-Rittenhouse, "Sex and the Single Vampire: The Evolution of the Vampire Lothario and Its Representation in *Buffy*," in *Fighting the Forces: What's at Stake in Buffy the Vampire Slayer*, eds. Rhonda V. Wilcox and David Lavery (New York: Rowman &Littlefield, 2002).
10. Beth Braun, "The X-Files and Buffy the Vampire Slayer: The Ambiguity of Evil in Supernatural Representations," *Journal of Popular Film and Television* 28, no. 2 (2000): 88–94; Jacqueline Reid-Walsh, "Power Girl/Girl Power: The Female Action Hero Goes to High School," *Journal of Adolescent and Adult Literacy* 42, no. 6 (1999): 503.
11. Modelski, *Loving with a Vengeance*.
12. Jeff Jensen, "Let 'Em Eat Stake," *Entertainment Weekly's EW.Com* [on-line] 11 June 2002. Available from World Wide Web: (http://www.ew.com/report/0,6115,260274~2|8971~,00.html).
13. *Spectrum Magazine* 1, August 2001, 15–16.
14. Ace, "Wrecked," *Mightybigtv.com* [on-line] 7 December 2001. Available from World Wide Web: (http://www.mightybigtv.com/story.cgi?show=12&story=2564&page=5&sort=&limit=).
15. *TVGuide.com* [on-line] 13 February 2002. Available from World Wide Web: (http://www.tvguide.com/).
16. "Buffy the Vampire Slayer Valentine's Giveaway," *Zap2it* [on-line] 6 February 2002. Available from World Wide Web: (http://tvzap2it.com/shows/features).
17. Stephanie Zacharek, "Buffy's Will-to-Power," *Salon.com* [on-line] 28 November 2001. Available from World Wide Web: (http://dir.salon.com/sex/feature/2001/11/28/buffy/index.html).
18. Dawn Heinecken, "Changing Ideologies in Romance Fiction," in *Romantic Conventions*, eds. Anne K. Kaler and Rosemary Johnson-Kurek (Bowling Green: Bowling Green State University Popular Press, 1999): 169.
19. Rhonda V. Wilcox, "Every Night I Save You": Buffy, Spike, Sex and Redemption," in *Slayage: The On-Line Journal of Buffy Studies* 5, 2002. [on-line] 15 August 2002. Available from World Wide Web: (http://www.slayage.tv/essays/wilcox.htm).
20. Dawn Heinecken, "Fan Readings of Sex and Violence on *Buffy the Vampire Slayer*" (paper presented at the Commonwealth Center for the Humanities and Society, Louisville, KY. April 2002).
21. Wilcox, "Every Night I Save You."
22. Wilcox, "Every Night I Save You."
23. Foucault, *Discipline and Punish*.
24. Carol Gilligan, *In a Different Voice: Psychological Theory and Women's Development* (Cambridge, MA: Harvard University Press, 1982): 105.
25. Ibid.

26. Rhonda V. Wilcox, "Who Died and Made Her the Boss? Patterns of Mortality in *Buffy*," in *Fighting the Forces: What's at Stake in Buffy the Vampire Slayer,* eds. Rhonda V. Wilcox and David Lavery (New York: Rowman & Littlefield, 2002).
27. Joss Whedon, "Interview, Welcome to the Hellmouth" [video], cited by Kathleen McDonnell.
28. It is not surprising that one of the recurring motifs of the season is rape, the ultimate denial of another's wishes. Buffy, for example, is attacked by a monster whose head is a penis in "Doublemeat Palace." She is immobilized by its bite and can't fight back—it is up to Willow, the lesbian, to chop the monster's head. Willow twice mind-rapes her friends and loved ones. The villainous "trio" brainwash a girl in an attempted rape that ends in her death.
29. Elyce Rae Helford, "My Emotions Give Me Power": The Containment of Girls' Anger in Buffy," in *Fighting the Forces: What's at Stake in Buffy the Vampire Slayer*, eds. Rhonda V. Wilcox and David Lavery (New York: Rowman & Littlefield, 2002): 18–35.
30. The visual imagery of this final episode reflects interesting notions about gender and power as well. Willow raises a satanic temple dedicated to a goddess. The temple itself is phallic, thrusting up from the ground, but is decorated with the figure of a woman adorned with snakes. The biblical imagery is reinforced by Xander who reminds us that he is a carpenter before he allows Willow to magically cut him over and over. Christ-like, he bleeds for her sins.
31. Wilcox, "Who Died and Made Her the Boss?" 8.
32. Diane DeKelb-Rittenhouse, "Sex and the Single Vampire: The Evolution of the Vampire Lothario and Its Representation in *Buffy*," in *Fighting the Forces: What's at Stake in Buffy the Vampire Slayer,* eds. Rhonda V. Wilcox and David Lavery (New York: Rowman & Littlefield, 2002): 147.
33. Gloria Goodale, "Television's Superwomen," *Christian Science Monitor*, 5 February 1999; and Micol Ostow, "Why I Love Buffy," *Sojourner* 24, no. 3 (1998); and Jennifer Pozner, "Thwack! Pow! Yikes! Not Your Mother's Heroines," *Sojourner* 23, no.12 (1997).
34. Helford, "My Emotions Give Me Power."
35. Lynne Edwards, "Slaying in Black and White: Kendra as Tragic Mulatta in *Buffy*," in *Fighting the Forces: What's at Stake in Buffy the Vampire Slayer*, eds., Rhonda V. Wilcox and David Lavery (New York: Rowman & Littlefield, 2002); and Donna Minkowitz, "Vampiros Lesbos," *Salon.com* [on-line] 12 January 2000. Available from World Wide Web: (http://www.salon.com/ent/log/2000/01/12/willow/index.html); and Ono, Kent A., "To Be a Vampire on *Buffy the Vampire Slayer*: Race and "Other" Socially Marginalizing Positions on Horror TV," in *Fantasy Girls: Gender in the New Universe of Science Fiction and Fantasy Televsion*, ed., Elyce Rae Helford (New York: Rowman & Littlefield, 2000): 163–186; and A. Susan Owen, "*Buffy the Vampire Slayer*: Vampires, Postmodernity and Postfeminism, " *Journal of Popular Film and Television* 27, no. 2 (1999): 31.

Chapter Six: Female Heroes and the Body in Postmodern Culture

1. Balsamo, "Forms of Technological Embodiment," 218.
2. Bryan S. Turner, *The Body and Society: Explorations in Social Theory* 2nd ed. (London: Sage, 1996): 19.
3. Turner, *The Body and Society,* 14.

4. Ibid., 14.
5. Ibid., 24.
6. Mike Featherstone, "The Body in Consumer Culture," in *The Body: Social Process and Cultural Theory,* eds. Mike Featherstone, Mike Hepworth, and Bryan S. Turner (London: Sage, 1991): 176.
7. Turner, *The Body and Society*, 19.
8. Ibid., 14.
9. Ibid., 20.
10. Lentz, "The Popular Pleasures of Female Revenge," 377.
11. Anne Balsamo, *Technologies of the Gendered Body: Reading Cyborg Women* (Durham: Duke University Press, 1996): 28.
12. John Cawelti and Bruce A. Rosenberg, *The Spy Story* (Chicago: University of Chicago Press, 1987): 13–14.
13. Ibid., 29.
14. Carol Clover, *Men, Women, and Chainsaws* (Princeton, NJ: Princeton University Press, 1992): 60.
15. Balsamo, *Technologies of the Gendered Body*, 30.
16. Ibid.
17. Mikhail Bakhtin, *Rabelais and His World*, trans. Helene Iswoldsky (Bloomington: Indiana University Press, 1981): 240, quoted in Rowe, *The Unruly Woman*, 34.
18. Ibid., 33–34.
19. Tom Carson, "La Femme Dilbert," *The Village Voice*, 13 January 1998, 77.
20. Kirsten Marthe Lentz, "The Popular Pleasures of Female Revenge (Or Rage Bursting in a Blaze of Gunfire)," *Cultural Studies* 7, no. 3 (1993): 381.
21. Susan J. Douglas, "Girls 'n' Spice: All Things Nice?" *The Nation*, 25 August–1 September 1997, 23.
22. Catharine Lumby, *Bad Girls: The Media, Sex and Feminism in the 90's* (St. Leonards, NSW: Allen & Unwin, 1997): 130.
23. Ibid.
24. Lumby, *Bad Girls*, 131.
25. Ibid.
26. Ibid., 134–135.
27. The allegiance to the conventions of "feminine" genres like soaps is consciously acknowledged by the WB Network, home of *Buffy*, *Roswell*, and *Angel*. A recent WB Network promo features a young man mimicking the romantic lines from WB series like *Dawson's Creek* in order to woo his girlfriend. The ad suggests that young men need to watch WB broadcasts in order to be clued in to the needs and desires of female audiences.
28. S. Elizabeth Bird, *For Enquiring Minds: A Cultural Study of Supermarket Tabloids* (Knoxville: University of Tennessee Press, 1992).
29. Lumby, *Bad Girls*, 132.
30. The same negation of death as an end of identity occurs in *Xena: Warrior Princess*. The characters make frequent trips to Hades and even return from the dead. At the end of the fourth season, Xena and Gabrielle are crucified and appear to be making a journey into the afterlife together.
31. Soren Kierkegaard, *The Concept of Dread*, cited by Adi Shmueli. *Kierkegaard and Consciousness*, trans. Naomi Handelman (Princeton, NJ: Princeton University Press, 1985): 80.

32. Linda Badley, "The Rebirth of the Clinic: The Body as Alien in *The X-Files*," in *Deny All Knowledge: Reading the X-Files*, ed. David Lavery, Angela Hague, and Maria Cartwright (Syracuse: Syracuse University Press, 1996): 163.
33. Ibid., 164.
34. Ibid., 166.
35. Featherstone, "The Body in Consumer Culture," 187.
36. Joan Jacobs Brumberg, *The Body Project: An Intimate History of American Girls* (New York: Random House, 1997).
37. David B. Morris, *Illness and Culture in the Postmodern Age* (Berkeley: University of California Press, 1998): 77.
38. Ibid., 221.
39. Ibid., 21–23.
40. Linda Badley, *Writing Horror and the Body: The Fiction of Stephen King, Clive Barker, and Anne Rice* (Westport, CT: Greenwood Press, 1996): 105.
41. Ibid., 116.
42. Ibid., 118.
43. Ibid., 129.

Epilogue: Female Heroes and Postfeminism

1. Patricia Pender, "I'm Buffy and You're…History": The Postmodern Politics of *Buffy*," in *Fighting the Forces: What's at Stake in Buffy the Vampire Slayer*, eds. Rhonda V. Wilcox and David Lavery (New York: Rowman & Littlefield, 2002): 43.
2. Green, *Cracks in the Pedestal*, 182.
3. Ibid., 183.
4. I don't think these series are unique. Superficially, Xena is less vulnerable than either Buffy or Nikita; she sets out to be a hero of her own free will. She's happier in her day-to-day work, and even has a female sidekick. But she is also an embattled character. Depicted as a woman constantly in conflict with herself, her internal world is her greatest threat. As Sherrie Inness writes, Xena's chief physical adversary, Callisto, is the physical embodiment of her evil side; Callisto became evil when Xena destroyed her village. Inness sees this as meaning that Xena acknowledges that power has a sinister side. I see it somewhat differently: In fighting Callisto, Xena must constantly face the monster to which she has metaphorically given birth (her sidekick, Gabrielle, literally gives birth to, and must fight, her demon daughter). Again, the body is the source and site of danger. Yet it is her danger, her evil, that gives her strength: She learned to fight as an evil warrior. If anything, the feminine but destructive power of the body in Xena is more traditional than in either Buffy or Nikita.
5. It is notable that other non-action shows like *Ally McBeal* and *Providence* feature hapless heroes who similarly struggle to maintain or develop a sense of self. The difference between the two genres is that in cloying shows such as *Ally McBeal*, the hero's lack of solid identity is accepted, normalized, and even applauded, while action genres at least tend to connect questions of identity to the problematic environment through which the hero must navigate.
6. Frances Early, "Staking Her Claim: *Buffy the Vampire Slayer* as Transgressive Woman Warrior," *Journal of Popular Culture* 35, no. 3 (2001): 22, 11–27.
7. The radio version of *Loveline* goes even further. The hosts actually bet on the kinds of past victimization experienced by the (usually female) caller. While certainly highlight-

ing a psychological tendency that people who have been victimized may be vulnerable to further abuse, this technique does little to acknowledge the fact that it is men who are doing the abuse.

8. Jean Baudrillard, *Simulacra and Simulation,* trans. Sheila Faria Glaser (Ann Arbor: University of Michigan Press, 1994).
9. Melissa R. Lafrance, "Colonizing the Feminine: Nike's Intersections of Postfeminism and Hyperconsumption," in *Sport and Postmodern Times*, ed. Genevieve Rail (Albany: State University of New York, 1998): 121.
10. Anna Quindlen, "And Now, Babe Feminism," in *Bad Girls/Good Girls: Women, Sex and Power in the Nineties*, eds. Nan Bauer Maglin and Donna Perry (New Brunswick, NJ: Rutgers University Press, 1996): 4.
11. Quindlen, "And Now, Babe Feminism," 4.

BIBLIOGRAPHY

Ace. "Wrecked." Mightybigtv.com [on-line] 7 December 2001. Available from World Wide Web: (http://www.mightybigtv.com/story.cgi?show=12&story=2564&page=5&sort=&limit=).

Aeon Flux: Mission Infinite. New York: MTV Network/Sony Music Entertainment, Inc./ Viacom International, 1997. Videocassette.

Aeon Flux: Operative Terminus. New York: MTV Network/ Sony Music Entertainment, Inc./ Viacom International, 1998. Videocassette.

Aeon Flux: The Video Game. Sony Playstation, December 1996; PC Version available from Cryo/Viacom New Media, January 1997.

Alexander, Max, Robin Honig, and Ethan Alter. "50 Greatest Shows of All Time." *TV Guide,* 4–10 May 2002, 18–55.

Ang, Ien. *Watching Dallas: Soap Opera and the Melodramatic Imagination*. London: Methuen, 1985.

Asher, Japhet, and Peter Gaffney. *The Herodotus File*. New York: MTV Books/Pocket Books, 1995.

Atkinson, Michael. "Inside Aeon." *Film Comment*, 1 January 1996, 20.

Badley, Linda. "The Rebirth of the Clinic: The Body as Alien in *The X-Files*." In *Deny All Knowledge: Reading The X-Files*, edited by David Lavery, Angela Hague, and Maria Cartwright. Syracuse, NY: Syracuse University Press, 1996.

Badley, Linda. *Writing Horror and the Body: The Fiction of Stephen King, Clive Barker, and Anne Rice*. Westport, CT: Greenwood Press, 1996.

Bakhtin Mikhail. *Rabelais and His World*. Translated by Helene Iswoldsky. Bloomington: Indiana University Press, 1981. Quoted in Kathleen Rowe, *The Unruly Woman: Gender and the Genres of Laughter* (Austin: University of Texas Press, 1995): 34.

Baldwin, Kristen. "Green's Day." *Entertainment Weekly*, 14 May 1999, 35–36.

Balsamo, Anne. "Forms of Technological Embodiment: Reading the Body in Contemporary Culture." In *Cyberspace, Cyberbodies, Cyberpunk: Cultures of Technological Embodiment*, edited by Mike Featherstone and Roger Burrows, 215–237. London: Sage, 1995.

Balsamo, Anne. *Technologies of the Gendered Body: Reading Cyborg Women*. Durham: Duke University Press, 1996.

Baudrillard, Jean. *Simulacra and Simulation* trans. Sheila Faria Glaser. Ann Arbor: University of Michigan Press, 1994.

Bellafante, Ginia. "Feminism: It's All About Me!" *Time*, 29 June 1998, 54–62.

Berger, John. *Ways of Seeing*. London: Penguin, 1977.

Bernstein, Jonathon. "Cherchez la Femme." *Gear*, September - October 1998, 72–79.

Bird, S. Elizabeth. *For Enquiring Minds: A Cultural Study of Supermarket Tabloids*. Knoxville: University of Tennessee Press, 1992.

Bordo, Susan. *Unbearable Weight: Feminism, Western Culture and the Body*. Berkeley: University of California Press, 1993.

Braun, Beth. "*The X-Files* and *Buffy the Vampire Slayer*: The Ambiguity of Evil in Supernatural Representations." *Journal of Popular Film and Television* 28, no. 2 (2000): 88–94.

Brumberg, Joan Jacobs. *The Body Project: An Intimate History of American Girls*. New York: Random House, 1997.

Brunsdon, Charlotte, Julie D'Acci, and Lynn Spigel. *Feminist Television Criticism: A Reader*. Oxford: Clarendon Press, 1997.

Buffy the Vampire Slayer: The Origin. Reading, MA: Federation Comics, 1999.

"Buffy the Vampire Slayer Valentine's Giveaway," *Zap2it* [on-line] 6 February 2002. Available from World Wide Web: (http://tvzap2it.com/shows/features)

Byars, Jackie. *All That Hollywood Allows: Re-Reading Gender in 1950's Melodrama*. Chapel Hill: University of North Carolina Press, 1991.

Carson, Tom. "La Femme Dilbert." *The Village Voice*, 13 January 1998, 77.

Cawelti, John. *Adventure, Mystery and Romance*. Chicago: The University of Chicago Press, 1976.

Cawelti, John, and Bruce A Rosenberg. *The Spy Story*. Chicago: The University of Chicago Press, 1987.

Chung, Peter. Message posted to *AOL's Aeon Flux Message Board*. [on-line] 24 September 1995. Available from World Wide Web: (http://members.aol.com/lalinton/from.htm).

Clover, Carol. *Men, Women, and Chainsaws*. Princeton, NJ: Princeton University Press, 1992.

Collins, Jim. *Uncommon Cultures: Popular Culture and Post-Modernism*. New York: Routledge, 1989.

Condit, Celeste. "The Rhetorical Limits of Polysemy." *Critical Studies in Mass Communication* 6 (1989): 103–122.

D'Acci, Julie. *Defining Women: The Case of Cagney and Lacey*. Chapel Hill: University of North Carolina Press, 1994.

Debord, Guy. *Society as Spectacle*. Detroit, MI: Black and Red Press, 1984.

Decandido, Keith R. A. *Buffy the Vampire Slayer: The Xander Years*. New York: Pocket Books, 1999.

DeKelb-Rittenhouse, Diane. "Sex and the Single Vampire: The Evolution of the Vampire Lothario and Its Representation in *Buffy*." In *Fighting the Forces: What's at Stake in Buffy the Vampire Slayer*, edited by Rhonda V. Wilcox and David Lavery, 143–152. New York: Rowman & Littlefield, 2002.

Douglas, Susan J. "Girls 'n' Spice: All Things Nice?" *The Nation*, 25 August-1 September 1997, 21–24.

Dow, Bonnie. *Prime-Time Feminism: Television, Media Culture, and the Women's Movement since 1970*. Philadelphia: University of Pennsylvania Press, 1996.

Early, Frances. "Staking Her Claim: *Buffy the Vampire Slayer* as Transgressive Woman Warrior." *Journal of Popular Culture* 35, no. 3 (2001): 11–27.

Edwards, Lynne. "Slaying in Black and White: Kendra as Tragic Mulatta in *Buffy*." In *Fighting the Forces: What's at Stake in Buffy the Vampire Slayer,* edited by Rhonda V. Wilcox and David Lavery, 85–97. New York: Rowman & Littlefield, 2002.

Fairclough, Norman. *Language and Power.* New York, Longman, 1989.

Featherstone, Mike. "The Body in Consumer Culture." In *The Body: Social Process and Cultural Theory,* edited by Mike Featherstone, Mike Hepworth and Bryan S. Turner, 170–196. London: Sage, 1991.

Ferrante, Anthony C. "Angel Flies Again." *Sci-Fi Teen*, November 1993, 37.

Fiske, John. *Television Culture*. London: Methuen, 1987.

Fiske, John, and John Hartley. "Bardic Television." In *Television: The Critical View*, edited by H. Newcomb, 600–612. New York: Oxford, 1987.

Flitterman-Lewis, Sandy. "Psychoanalysis, Film, and Television." In *Channels of Discourse, Reassembled: Television and Contemporary Criticism*. 2nd ed., edited by Robert C. Allen, 203–246. Chapel Hill: University of North Carolina Press, 1992.

Foucault, Michel. *Discipline and Punish: The Birth of the Prison*. Trans. Alan Sheridan. New York: Random House, 1995.

Gardner, Craig Shaw. *Return to Chaos*. New York: Pocket Books, 1998.

Gilligan, Carol. *In a Different Voice: Psychological Theory and Women's Development*. Cambridge, MA: Harvard University Press, 1982.

Gledhill, Christine. "Developments in Feminist Film Criticism." In *Re-Visions: Essays in Feminist Film Criticism*, edited by Mary Ann Doane, Patricia Mellencamp, and Linda Williams, 18–48. Frederick, MD: American Film Institute and University Press of America, 1984.

Gledhill, Christine. "Pleasurable Negotiations." In *Female Spectators: Looking at Film and Television*, edited by E. Deidre Pribram, 64-89. London: Verso, 1988.

Goldblatt, Henry. "The Universal Appeal of Schlock." *Fortune*, 12 May 1997, 32.

Goldblatt, Henry. *Variety*, 17 March 1997, 27.

Goldblatt, Henry. *Variety,* 17 February 1997, 39.

Goodale, Gloria."Television's Superwomen," *Christian Science Monitor*, 5 February 1999.

Graham, Jefferson. "Nikita's Star Shoots Straight for the Top: Peta Wilson Knocks' Em Dead on Cable." *USA Today*, 14 April 1997.

Green, Philip. *Cracks in the Pedestal: Ideology and Gender in Hollywood.* Amherst: University of Massachusetts Press, 1998.

Hadleigh-West, Maggie. *War Zone*. Northampton, MA: Media Education Foundation, 1998. Videocassette.

Hall, M.A. *Feminism and Sporting Bodies: Essays on Theory and Practice*. Champaign, IL: Human Kinetics, 1996.

Haskell, Molly. *From Reverence to Rape: The Treatment of Women in the Movies*. New York: Holt, Rinehart and Winston, 1973.

Heinecken, Dawn. "Changing Ideologies in Romance Fiction." In *Romantic Conventions*, edited by Anne K. Kaler and Rosemary Johnson-Kurek. Bowling Green: Bowling Green State University Popular Press, 1999.

Heinecken, Dawn. "Fan Readings of Sex and Violence on *Buffy the Vampire Slayer*." Paper presented at the Commonwealth Center for the Humanities and Society, Louisville, KY. April 2002.

Helford, Elyce Rae. "My Emotions Give Me Power": The Containment of Girls' Anger in Buffy." In *Fighting the Forces: What's at Stake in Buffy the Vampire Slayer*, edited by Rhonda V. Wilcox and David Lavery, 18–35. New York: Rowman & Littlefield, 2002.

Inness, Sherrie A. *Tough Girls: Women Warriors and Wonder Women in Popular Culture*. Philadelphia: University of Pennsylvania Press, 1999.

Jenkins, Henry. *Textual Poachers: Television Fans and Participatory Culture*. New York: Routledge, 1992.

Jensen, Jeff. "Let 'Em Eat Stake." *Entertainment Weekly's EW.Com*, 2002. [on-line] 11 June 2002. Available from World Wide Web: (http://www.ew.com/report/0,6115,260274~2|8971~,00.html).

Kellner, Douglas. *Media Culture: Cultural Studies, Identity and Politics Between the Modern and Postmodern*. New York: Routledge, 1995.

La Femme Nikita: Music from the Television Series. New York: TVT Records, 1998.

Lafrance, Melissa R. "Colonizing the Feminine: Nike's Intersections of Postfeminism and Hyperconsumption." In *Sport and Postmodern Times*, edited by Genevieve Rail, 117–139. Albany: State University of New York Press, 1998.

Lentz, Kirsten Marthe. "The Popular Pleasures of Female Revenge (Or Rage Bursting in a Blaze of Gunfire)." *Cultural Studies* 7, no. 3 (1993): 374–405.

Lumby, Catharine. *Bad Girls: The Media, Sex and Feminism in the 90's*. St. Leonards, NSW: Allen & Unwin, 1997.

McClary, Susan. *Feminine Endings: Music, Gender and Sexuality*. Minneapolis: University of Minnesota Press, 1991.

Mellencamp, Patricia. *High Anxiety: Catastrophe, Scandal, Age, and Comedy*. Bloomington: Indiana University Press, 1992.

Messner, Michael A. "Sports and Male Domination: The Female Athlete as Contested Ideological Terrain." In *Philosophic Inquiry in Sport*. 2nd ed., edited by William Morgan and Klaus Meier, 274–283. Champaign, IL: Human Kinetics, 1995.

Minkowitz, Donna. "Vampiros Lesbos." *Salon.com,* 12 January 2000. [on-line] 15 August 2002. Available from World Wide Web: (http:// www.salon.com/ent/log/2000/01/12/willow/index.html).

Modleski, Tania. *Loving with a Vengeance: Mass-Produced Fantasies for Women*. New York: Routledge, 1982.

Modleski, Tania. *Studies in Entertainment*. Bloomington: Indiana University Press, 1986.

Morley, David. *The 'Nationwide' Audience: Structure and Decoding*. London: British Film Institute, 1980.

Morris, David B. *Illness and Culture in the Postmodern Age*. Berkeley: University of California Press, 1998.

Mulvey, Laura. "Visual Pleasure and Narrative Cinema." In *Contemporary Film Theory*, edited by Antony Easthope. New York: Longman, 1993.

"New AFI Awards Show Airs Saturday." *CNN.com*, 4 January 2002 [on-line] 15 August 2002. Available from World Wide Web: (http:// www.cnn.com/2002/SHOWBIZ/News/01/04awards.afi.advance).

News Brief. *SciFi Wire: A News Service of the Sci Fi Channel*, 28 June 1999. [on-line] 28 June 1999. Available from World Wide Web: (http://www.scifi.com).

neXus siX magaZine, 1995/1996. [on-line] 15 May 1999. Available from World Wide Web: (http:// www.hallucinet.com/n6/aeon.htm).

Notebook. *Time Magazine*, 22 July 1996, 19.

Ono, Kent A. "To Be a Vampire on *Buffy the Vampire Slayer*: Race and "Other" Socially Marginalizing Positions on Horror TV." In *Fantasy Girls: Gender in the New Universe of Science Fiction and Fantasy Televsion*, edited by Elyce Rae Helford, 163–186. New York: Rowman & Littlefield, 2000.

"The O'Reilly Factor." Fox. 26 April 1999.

O'Shea, Tara. "The Cat." [on-line] 6 April 1999. Available from World Wide Web: (http://www.geocities.com/-teamhair/cat.html).

Ostow, Micol. "Why I Love Buffy." *Sojourner* 24, no. 3 (1998).

Owen, A. Susan. "Buffy the Vampire Slayer: Vampires, Postmodernity and Postfeminism. " *Journal of Popular Film and Television* 27, no. 2 (1999): 24–31.

Pender, Patricia. "I'm Buffy and You're...History": The Postmodern Politics of *Buffy*." In *Fighting the Forces: What's at Stake in Buffy the Vampire Slayer*, edited by Rhonda V. Wilcox and David Lavery, 35–44. New York: Rowman & Littlefield, 2002.

Pini, Maria. "Women and the Early British Rave Scene." In *Back to Reality? Social Experience and Cultural Studies*, edited by Angela McRobbie, 152–169. Manchester: Manchester University Press, 1997.

Pozner, Jennifer. "Thwack! Pow! Yikes! Not Your Mother's Heroines." *Sojourner* 23, no.12 (1997).

Quindlen, Anna. "And Now, Babe Feminism." In *Bad Girls/Good Girls: Women, Sex and Power in the Nineties*, edited by Nan Bauer Maglin and Donna Perry, 3–5. New Brunswick, NJ: Rutgers University Press, 1996.

Radway, Janice. *Reading the Romance: Women, Patriarchy and Popular Literature*. Chapel Hill: University of North Carolina Press, 1984.

Reid-Walsh, Jacqueline. "Power Girl/Girl Power: The Female Action Hero Goes to High School." *Journal of Adolescent and Adult Literacy* 42, no. 6 (1999): 502–503.

Rosen, Marjorie. *Popcorn Venus: Women, Movies, and the American Dream*. New York: Coward, McCann & Geohage, 1973.

Roush, Matt. "The Roush Review." *TV Guide*, 28 March 28–3 April 1998, 9.

Rowe, Kathleen. *The Unruly Woman: Gender and the Genres of Laughter*. Austin: University of Texas Press, 1995.

Rowe, Kathleen. "Roseanne: Unruly Woman as Domestic Goddess." In *Feminist Television Criticism: A Reader*, edited by Charlotte Brunsdon, Julie D'Acci and Lynn Spigel, 74–83. Oxford: Clarendon Press, 1997.

Rubin, Sylvia. "Back for More/The New Femme Fatale: USA's Popular 'Nikita' an Assassin with Sizzle." *San Francisco Chronicle*, 17 September 1997.

Segal, Lynn. *Slow Motion: Changing Masculinities, Changing Men*. New Brunswick, NJ: Rutgers University Press, 1990.

Seiter, Ellen, Hans Borchers, Gabriele Kreutzer, and Eva-Maria Warth. "Introduction." In *Remote Control: Television, Audiences and Cultural Power*, edited by Ellen Seiter, Hans Borchers, Gabriele Kreutzer, and Eva-Maria Warth, 1–15. London: Routledge, 1989.

Shields, Vickie Rutledge, and Dawn Heinecken. *Measuring Up: How Advertising Affects Self-Image*. Philadelphia: University of Pennsylvania Press, 2002.

Shmueli, Adi. *Kierkegaard and Consciousness*. Trans. Naomi Handelman. Princeton, NJ: Princeton University Press, 1985.

Silverman, Kaja. *The Subject of Semiotics*. Oxford: Oxford University Press, 1983.

Spectrum Magazine 1, August 2001, 15–16.

Spitzack, Carole. *Confessing Excess: Women and the Politics of Body Reduction*. Albany: State University of New York Press, 1990.

Stacey, Jackie. *Star Gazing: Hollywood Cinema and Female Spectatorship*. New York: Routledge, 1994.

Tankersley, Richie. *Buffy the Vampire Slayer*. New York: Pocket Books, 1997.

Tankersley, Richie. *The Harvest*. New York: Pocket Books, 1997.

Tankersley, Richie. *The Angel Chronicles*. New York: Archway, 1999.

Tasker, Yvonne. *Working Girls: Gender and Sexuality in Popular Cinema*. New York: Routledge, 1998.

Tasker, Yvonne. *Spectacular Bodies: Gender, Genre, and the Action Cinema*. New York: Routledge, 1993.

Tasker, Yvonne. "Fists of Fury: Discourses of Race and Masculinity in the Martial Arts Cinema." In *Race and the Subject of Masculinities*, edited by Harry Stecopoulos and Michael Uebel, 315–336. Durham: Duke University Press, 1997.

Tompkins, Jane. *West of Everything: The Inner Life of Westerns*. New York: Oxford University Press, 1992.

Traube, Elizabeth G. *Dreaming Identities: Class, Gender and Generation in 1980's Hollywood Movies*. Boulder: Westview Press, 1992.

Turner, Bryan S. *The Body and Society: Explorations in Social Theory*. 2nd ed. London: Sage, 1996.

TVGuide.com. [on-line] 13 February 2002. Available from World Wide Web:(http://tvguide.com)

Warner, William. "Spectacular Action: Rambo and the Popular Pleasures of Pain." In *Cultural Studies*, edited by Lawrence Grossberg, Cary Nelson, and Paula Treichler, 672–688. New York: Routledge, 1991.

Whedon, Joss. "Interview." *Welcome to the Hellmouth* [video] Cited by Kathleen McDonnell, "Chaos at the Mouth of Hell." *Gothic Studies* 2, no. 1 (2000).

Whitehead, Colson. "That Girl." *The Village Voice*, 22 August 1995, 43.

Wilcox, Rhonda V. "Every Night I Save You": Buffy, Spike, Sex and Redemption." In *Slayage: The On-Line Journal of Buffy Studies* 5, 2002. [on-line] 21 August 2002. Available from World Wide Web: (http://www.slayage.tv/essays/wilcox.htm).

Wilcox, Rhonda V. "Who Died and Made Her the Boss? Patterns of Mortality in *Buffy*." In *Fighting the Forces: What's at Stake in Buffy the Vampire Slayer*, edited by Rhonda V. Wilcox and David Lavery, 3–17. New York: Rowman & Littlefield, 2002.

Williams, Linda. "Feminist Film Theory: *Mildred Pierce* and the Second World War." In *Female Spectators: Looking at Film and Television*, edited by E. Diedre Pribram, 12–30. London: Verso, 1988.

Williams, Linda. "'Something Else Besides a Mother': *Stella Dallas* and the Maternal Melodrama." In *Issues in Feminist Film Critiicism*, edited by Patricia Erens, 137–162. Bloomington: Indiana University Press, 1990.

Willis, Susan. *A Primer for Daily Life*. New York: Routledge, 1991.

Young, Iris Marion. *Throwing like a Girl and Other Essays in Feminist Philosophy and Social Theory*. Bloomington: Indiana University Press, 1990.

Zacharek, Stephanie. "Buffy's Will-to-Power." *Salon.com* [on-line] 28 November 2001. Available from World Wide Web: (http://dir.salon.com/sex/feature/2001/11/28/buffy /index.html).

Zacharek, Stephanie."Willow, Destroyer of Worlds." *Salon.com* [on-line] 30 May 2002. Available from World Wide Web: (http://www.salon.com/ent/letters/2002/05/30/buffy/ index.html).

INDEX

C

D

E

N

O

W

X

Y

Z

Intersections in Communications and Culture

Global Approaches and Transdisciplinary Perspectives

General Editors: Cameron McCarthy & Angharad N. Valdivia

An Institute of Communications Research, University of Illinois Commemorative Series

This series aims to publish a range of new critical scholarship that seeks to engage and transcend the disciplinary isolationism and genre confinement that now characterizes so much of contemporary research in communication studies and related fields. The editors are particularly interested in manuscripts that address the broad intersections, movement, and hybrid trajectories that currently define the encounters between human groups in modern institutions and societies and the way these dynamic intersections are coded and represented in contemporary popular cultural forms and in the organization of knowledge. Works that emphasize methodological nuance, texture and dialogue across traditions and disciplines (communications, feminist studies, area and ethnic studies, arts, humanities, sciences, education, philosophy, etc.) and that engage the dynamics of variation, diversity and discontinuity in the local and international settings are strongly encouraged.

LIST OF TOPICS

- Multidisciplinary Media Studies
- Cultural Studies
- Gender, Race, & Class
- Postcolonialism
- Globalization
- Diaspora Studies
- Border Studies
- Popular Culture
- Art & Representation
- Body Politics
- Governing Practices
- Histories of the Present
- Health (Policy) Studies
- Space and Identity
- (Im)migration
- Global Ethnographies
- Public Intellectuals
- World Music
- Virtual Identity Studies
- Queer Theory
- Critical Multiculturalism

Manuscripts should be sent to:

Cameron McCarthy OR Angharad N. Valdivia
Institute of Communications Research
University of Illinois at Urbana-Champaign
222B Armory Bldg., 555 E. Armory Avenue
Champaign, IL 61820

To order other books in this series, please contact our Customer Service Department:

(800) 770-LANG (within the U.S.)
(212) 647-7706 (outside the U.S.)
(212) 647-7707 FAX

Or browse online by series:

www.peterlangusa.com